BEYOND BOUNDARIES

Amazon
2/14
$17

BEYOND BOUNDARIES

THE NEW NEUROSCIENCE OF CONNECTING BRAINS WITH MACHINES—AND HOW IT WILL CHANGE OUR LIVES

MIGUEL NICOLELIS

ST. MARTIN'S GRIFFIN
NEW YORK

Printed in the United States of America. For information, address St. Martin's Press, 175 Fifth Avenue, New York, N.Y. 10010.

www.stmartins.com

Design by Kelly S. Too

The Library of Congress has cataloged the Henry Holt edition as follows:

Nicolelis, Miguel.
Beyond boundaries : the new neuroscience of connecting brains with machines—and how it will change our lives / Miguel Nicolelis.—1st ed.
p. cm.
Includes bibliographical references and index.
ISBN 978-0-8050-9052-9
1. Thought and thinking. 2. Neurosciences. 3. Brain. 4. Machinery. I. Title.
BF441.N494 2011
003'.5—dc22 2010028440

ISBN 978-1-250-00261-7 (trade paperback)

Originally published in hardcover format by Times Books, an imprint of Henry Holt and Company

First St. Martin's Griffin Edition: March 2012

10 9 8 7 6 5 4 3 2 1

To Giselda and Ângelo,
For a lifetime of unconditional love

Be not the slave of your own past. Plunge into the sublime seas, dive deep and swim far, so you shall come back with self-respect, with new power, with an advanced experience that shall explain and overlook the old.

Ralph Waldo Emerson

CONTENTS

BEYOND BOUNDARIES

PROLOGUE: JUST FOLLOW THE MUSIC

As the first violin arpeggios emerged from the marble walls of the ample hall and capriciously ventured down the stairs from the second floor to the main entrance of the deserted medical school building, I could not help but feel disoriented by the total absurdity of the situation. After all, no medical student would be prepared to find himself listening to a concerto in the middle of the night while taking a quick break from one of the busiest hospital emergency rooms in the world. Yet, my initial uneasiness was soon replaced by music that breathed a whole new life, full of hope and adventure, into a soggy tropical summer evening. Perhaps that is why, even though those arpeggios seduced my brain nearly a quarter of a century ago, I can still vividly recall how the stunning beauty of the melody, not the otherwise meaningless individual notes, composed an earnest collective plea that beckoned me to follow the siren music. I swiftly bounded up the stairs and mutely walked through a thin corridor to find myself standing at the entrance to the auditorium where the "Vorspiel," the overture of Wagner's *Parsifal*, was remorselessly playing. Unable to resist, I followed the music and entered the auditorium.

How disappointing it felt, then, when I realized that, except for an elderly, well-dressed gentleman who was busily working, apparently trying to fix a worn-out faulty projector that had mangled one too many of his slides over the years, the auditorium, with all its chandeliers blazing, was completely empty. Built in the late 1920s, each of the classroom auditoriums at the University of São Paulo medical school was a model of elegant economy. At the front, a tidy, boxlike stage demarcated the space from which professors lectured. A heavy wood table, a sturdy chair, and a long, well-worn sliding blackboard completed the humble teacher's domain. The student seating was stacked into steep, straight rows, allowing backbenchers inhabiting the last row—including me—to live well beyond the *catedraticos*' authoritative stare during the unending lectures.

By now, the old man—his close-cropped white hair matching his pristine lab coat—was startled by the sound as I opened the door to the lecture hall. Yet he turned to me to reveal his effortless Mediterranean smile. Without giving up his struggle with the projector, he waved his left hand almost as if we had known each other for years. On the lecturer's desk I saw, to my dismay, the evidence that clearly incriminated that unsuspicious gentleman with that night's musical recital: a turntable, two expensive-looking loudspeakers, and the covers of a few records by the Berlin Philharmonic.

"Come in. Welcome. We have wine and cheese. I am having some difficulties with the projector tonight, but we will be ready to start in a moment. By the way, my name is Professor César Timo-Iaria. I am the teacher of this course."

He had barely finished the sentence when a loud metallic pinging resounded from the slide projector and light spilled onto the lecture hall's screen. Without waiting for my reply, he swiftly changed position to stand behind the projector, looking very much like a battle-proofed admiral on his ship's bridge. After dimming the chandeliers and waiting for the second track of the record to start playing, he began clicking through his slides with a joy that I had only seen and experienced as a child while playing soccer in the narrow streets of my old neighborhood. Sitting there alone in the dark, lullabied by Tannhäuser's singing, which echoed around the entire auditorium while images wholly unrelated to the typical medical curriculum were glancing off the screen, I felt both provoked and enticed like in no other lecture I had attended before in my life.

"But what course are you teaching?" I asked.

"Introduction to Physiology," Professor Timo-Iaria replied, without looking at me.

Just to be sure, I looked at the screen again. Like all medical students, I had taken the mandatory introduction to physiology course a few years earlier and, as far as I could tell, none of the images I was seeing matched what I had been taught then.

"How come?" I insisted.

"How come what, son?" he rebutted, still not looking at me.

"How could this be 'introduction to physiology'? Your slides, they are all about, I mean, you are showing only . . ."

"Yes?" He looked amused by my discomfort, as if he had seen this happen many times before. "Go ahead. Tell me what is so surprising to you."

The music, the images, an old man lecturing in the middle of the night in a vast and empty auditorium. Nothing made any sense. Half-perplexed and mildly irritated, I finally let him have it.

"Stars, galaxies, those are the images you are showing. Look, now there is a radio telescope on the screen. What is this? How could this be an introduction to physiology?"

"Well, this was the beginning. It all started there, from the big bang to brains in just about fifteen billion years. Quite a voyage, wasn't it? I will explain what I mean."

Through an endless visual parade of mindless shining spiral galaxies, budding star clusters, playful nebulas, rebellious comets, and exploding supernovas, all tendered by music that seemed to have been composed by universal gods, I watched Dr. Timo-Iaria's slide-by-slide depiction of the epic that led to the emergence of the human mind. Planets were formed. Most remained bare, lifeless lands. But on at least one, an interesting experiment led to the emergence, a few billion years ago, of the biochemical and genetic mechanisms for sustaining and replicating life. And life blossomed, struggled to survive, and, always full of hope and aspirations, started to evolve through many utterly unpredictable and tenuous roads.

Next, I saw images of the first hominid couples walking, side by side, millions of years ago in the middle of an African night in what today is Ethiopia's Afar Desert. And then, at the very moment when Wagner's

Tannhäuser was at last granted freedom from the Venusberg by rejecting immortality for the simple reward of experiencing what it is to be human, I shared the instant in which those early ancestors first looked at the infinite bright sky above them, full of awe and fear, while a raging electrical storm crisscrossed their brains searching for answers to the questions that still torment us today. I realized that by looking timidly but curiously to the sky, those first men and women launched a long and noble relay race that since has united us all in the search for the fundamental explanations of our existence, our consciousness, and the meaning of all that surrounds us. The symbolic birth of science could not have been better chronicled. Clearly, the seasoned admiral on the bridge knew very well how to steer his ship.

The dying notes of Tannhäuser's "Pilgrim's Chorus" announced the final slide, which, after being projected on the screen, lingered there as both of us remained in solemn silence. The slide showed a side view of a human brain. After a couple of minutes, Dr. Timo-Iaria turned on the lights, came down from his station by the projector, and walked calmly toward the auditorium door. Before leaving the room, he turned as if to say good-bye. Instead, he said: "This is the first lecture of the introduction to human physiology course. But I forgot to mention that I also teach an advanced course on neurophysiology. The first class is tomorrow night. I strongly advise you to take that class, too."

Stunned by what I had just experienced, I could only think of asking, "And what do I need to do to enroll in that course?"

Smiling again as he exited through the hall, Dr. Timo-Iaria dispensed his very first piece of advice to the lifelong student he had by now so effortlessly recruited.

"Just follow the music."

For the past twenty-five years, I have often remembered Dr. Timo-Iaria's unshakable belief that music and the scientific method represent two of the most astounding by-products to emerge from the endless toils and torments of the human mind. That may explain why I decided to dedicate my whole career to listening to a different type of music, the kind of symphonies composed by vast ensembles of brain cells.

Technically speaking, I am a systems neurophysiologist. At least this is the way most of my neuroscience colleagues would define the type of work that my students and I carry out in my laboratory at the Center for Neuroengineering at Duke University, in Durham, North Carolina. In general terms, systems neurophysiologists spend their lives investigating the physiological principles that underlie the operation of the large variety of neural circuits formed by nerve fibers that emanate from the hundreds of billions of cells that inhabit our brains. Such intricate brain networks, which dwarf by many orders of magnitude the complexity and connectivity of any electrical, computational, or machine grid ever assembled by humans, allow each individual brain cell, known as a neuron, to establish direct contact and communicate with hundreds or even thousands of its peers. By virtue of their particular morphology, neurons are highly specialized in receiving and transmitting minute electrochemical messages through their cellular contacts, called synapses, which they use to communicate with other neurons. It is through these immensely interconnected and highly dynamic cellular networks, which are known rather prosaically as neural circuits, that the brain goes about its main business: the production of a multitude of specialized behaviors that collectively define what we usually, and proudly, refer to as "human nature."

By harnessing massive waves of millivolt electrical discharges, these microscopic neural grids truly provide for every act of thinking, creation, destruction, discovery, cover-up, communication, conquest, seduction, surrender, love, hate, happiness, sadness, solidarity, selfishness, introspection, and elation ever perpetrated by every one of us, and our ancestors, throughout humanity's whole existence. Had the word *miracle* not been appropriated by another area of human enterprise, I believe society should grant neuroscientists exclusive rights to use it when reporting the wonders that brain circuits can generate on a routine basis.

For most systems neurophysiologists, like me, the ultimate quest is to decipher the physiological mechanisms that allow these bursts of neurobiological electricity to give birth to the vast repertoire of human action and behavior. In seeking that holy grail, however, much of neuroscience over the past two hundred years has been embroiled in settling the hotly contested dispute over what specific regions of the brain serve a particular function or behavior. At one extreme, radical localizationists, who are the legitimate but often unclaimed heirs of Franz Gall, the father of phrenology, still firmly believe that distinct brain functions are generated by highly specialized and spatially segregated areas of the nervous system. In the other corner, a smaller but fast-growing crowd, whom I call the distributionists, professes that rather than relying solely on unique specialization, the human brain calls on populations of multitasking neurons, distributed across multiple locations, to achieve every one of its goals. In defending this position, we distributionists propose that the brain seems to utilize a physiological mechanism that is somewhat equivalent to an election, a neuronal vote in which large populations of cells located in many different regions of the brain contribute, albeit each in small and different amounts, to the generation of a final behavioral product.

Over the past two centuries, both the localizationist and the distributionist camps have elected the cortex—the most superficial component of the brain, lying just beneath the bone layer of the skull—as the main neuronal battlefield for their never-ending dispute. The origins of this battle can be traced to the days when phrenologists claimed to be able to recognize key personality traits of an individual simply by palpating the scalp in search of skull bumps that reflected the disproportional enlargement of particular areas of the cortex that, according to their doctrine, generated attributes such as affection, pride, arrogance, vanity, and ambition. According to this doctrine, each human emotion and behavior was generated by a particular cortical territory.

Although Gall and his pseudoscience were discredited in due time, this general framework survived and morphed into one of the key dogmas of twentieth-century neuroscience. About one hundred years ago, a glorious series of experiments by the first generation of full-time brain researchers, led by the genial Spaniard Santiago Ramón y Cajal, demonstrated that, as in all other organs, an individual cell, the neuron, constitutes the brain's fundamental anatomical unit. Almost by default, the single neuron was also quickly anointed as the fundamental *functional* unit of the central nervous system. The ascension of the single neuron doctrine, combined with a dazzling 1861 report by Pierre Paul Broca, the French physician who observed that a localized lesion of the left frontal lobe could lead to a profound loss of speech and paralysis of the right half of a patient's body, temporarily put the distributionist camp in disarray. But just as the distributionists were becoming isolated, Sir Charles Sherrington came to their rescue. Sherrington argued that even one of the brain's simplest functions, the spinal cord arch reflex, depended on the collaboration of many neurons and distinct neural circuits to work properly.

In the last decade, although no decisive blow has yet been delivered, the distributionists have gained the high ground in the battle over the brain's soul. Discoveries emanating from neuroscience laboratories around the world are overturning the localizationists' model. Among these collective efforts, research conducted in my lab at Duke University over the last two decades has helped to show categorically that a single neuron can no longer be viewed as the fundamental functional unit of the brain; instead, connected populations of neurons are responsible for the symphonies of thought composed by brains. Today, we can record the music produced by these neural ensembles and even replay a small fraction of it in the form of concrete and voluntary motor behaviors. By listening to just a few hundred neurons—an infinitesimally small sample of the billions of neurons in the brain—we are already beginning to replicate the process by which complex thoughts become instantaneous body actions.

What principles guide the composition and conduction of these neural symphonies? After more than two decades delving into the workings of neural circuits, I have found myself looking for those principles both outside the brain, beyond the boundaries that have constrained our biological evolution out of humble beginnings in stardust, as well as deep inside the central nervous system, trying to identify and give voice to the *brain's own point of view.* Here I propose that, like the universe that fascinates us so much, the human brain is a relativistic sculptor; a skillful modeler that fuses neuronal space and time into an organic continuum responsible for creating all that we see and feel as reality, including our very sense of being. In the following chapters, I will propose that, in the next decades, by combining such a relativistic view of the brain with our growing technological ability to listen and decode even larger and more complex neural symphonies, neuroscience will eventually push human reach way beyond the current constraints imposed by our fragile primate bodies and sense of self.

I can imagine this world with some confidence because of the work conducted by my lab to teach monkeys to utilize a revolutionary neurophysiological paradigm, which we named brain-machine interfaces (BMIs). Using such BMIs, we were able to demonstrate that monkeys could learn to control voluntarily the movements of extraneous artificial devices, such as robotic arms and legs, located either close to or very far from them, using only their raw electrical brain activity. This unleashes a vast array of possibilities for the brain and the body that could, in the long run, completely change the way we go about our lives.

To test the different versions of our BMIs, we took advantage of a new experimental approach to read directly and simultaneously the electrical signals produced by hundreds of neurons that belong to a neural circuit. This technology was initially developed as a way to test the distributionists' viewpoint: that populations of single neurons, communicating with one another across different brain regions, are required to generate any brain function. But once we discovered how to listen to some motor neural symphonies played by the brain, we decided to push further: to record, decode, and transmit—all the way to the other side of the world—the motor thoughts of a primate cortex. We then translated these thoughts into digital commands to generate humanlike motion in machines that were never designed to acquire such unique human traits. It was at that moment that our BMIs stumbled on a way to liberate the brain from the constraints imposed by the body and made it capable of using virtual, electronic, and mechanical tools to control the physical world. Just by thinking. This book tells the story of those experiments and how they have changed our understanding of brain function.

For the vast majority of people alive today, the full impact of our research with BMIs will be felt primarily in the medical arena. Unraveling the brain's intricate workings by building advanced BMIs will lead to the development of amazing new therapies and cures for those afflicted by devastating neurological disorders. Such patients will be allowed to regain mobility and the sense and feeling in an otherwise lame body through a variety of neuroprosthetics, devices the size of a modern heart pacemaker that harvest healthy brain electrical activity to coordinate the contractions of a silk-thin wearable robot, a vest as delicate as a second skin but as protective as a beetle's exoskeleton—a suit capable of supporting a paralyzed person's weight and making formerly immobile bodies roam, run, and once again exult in exploring the world freely.

Yet, BMI applications promise to reach way beyond the borders of medicine. I believe future generations will be in a position to enact deeds and experience sensations that few today can imagine, let alone verbalize. BMIs may transform the way we interact with the tools we fabricate and how we communicate with one another and with remote environments and worlds. To grasp what this future world may look like, you first need to picture how the execution of a few of our daily routines will change radically when our brain's electrical activity acquires the means to roam freely around the world pretty much like radio waves sail above us today. For a moment, imagine living in a world where people use their computers, drive their cars, and communicate with one another simply by thinking. No need for cumbersome keyboards or hydraulic steering wheels. No point in relying on body movements or spoken language to express one's intentions to act upon the world.

In this new brain-centered world, such newly acquired neurophysiological abilities will seamlessly and effortlessly extend our motor, perceptual, and cognitive skills to the point that human thoughts can be efficiently and flawlessly translated into the motor commands needed to produce either the minute manipulations of a nanotool or the complex maneuvers of a sophisticated industrial robot. In that future, back at your beach house, sitting in your favorite chair facing your favorite ocean, you may one day effortlessly chat with any of a multitude of people anywhere in the world over the Internet without typing or uttering a single word. No muscle contraction involved. Just by thinking.

If that is not enticing enough, how about experiencing all the sensations aroused by touching the surface of a different planet, millions of miles away, without leaving your living room? Or even better, how would you feel if you could access your ancestral memory bank and readily download the thoughts of one of your forefathers and create, through his most intimate impressions and vivid memories, an encounter you both would have never shared otherwise? That is just a glimpse of what living in a world beyond the boundaries imposed upon the brain by the body may bring to our species.

Such wonders will soon no longer be the stuff of science fiction. This world is starting to take shape before our very eyes, right here and right now. And to become immersed in it, as Dr. Timo-Iaria would say, all you have to do is just follow the music that begins playing on the very next page.

1

WHAT IS THINKING?

By the time the rainy days of the tropical autumn of 1984 arrived, most Brazilians had had enough. For twenty years, their beloved country had been ruled by a vicious dictatorship, brought to power by a military coup d'état that triumphed, emblematically, on April Fools' Day 1964. For the next two decades the military regime built an infamous legacy, marked primarily by its rampant incompetence, widespread corruption, and shameful political violence against its own people.

By 1979, thanks to the growing popular opposition to the regime, the latest four-star general installed in the presidential palace had no alternative but to grant amnesty to the political leaders, scientists, and intellectuals who had fled into exile abroad. A gradual, controlled return to civilian rule had been mapped out by the generals, beginning with popular elections for state governorships in the fall of 1982.

That November, the opposition parties won by a landslide. By the next year, however, that small token of democracy had been all but forgotten. Brazilians realized they had the right and, more importantly, the power to demand more than a dictator's political bread crumbs. They wanted to oust the military government, but not through another coup d'état. Instead, they wanted to vote it into retirement through a direct election for president. That is how, seemingly out of nowhere, a nationwide movement demanding immediate direct elections for president (*diretas já* in Portuguese) broke loose. The first rally took place in the tiny northeastern city of Abreu e Lima on March 31, 1983. By November, a somewhat shy crowd of ten thousand people had gathered to protest in Brazil's most populous and wealthy city, São Paulo. From that point, the movement grew exponentially. Two months later, on January 25, 1984, the day São Paulo celebrated its 430th anniversary, more than two hundred thousand people were chanting their collective demand for immediate presidential elections. In a matter of days, gigantic crowds started to converge on the main squares of Rio de Janeiro, Brasilia, and other major cities.

On the evening of April 16, 1984, more than one million people congregated in the heart of downtown São Paulo to participate in the largest political rally ever staged in the country's history. In a matter of hours, a river of people, most dressed in the Brazilian national colors, green and yellow, inundated the valley where the city was originally founded. Every new group of people that arrived immediately joined into an already familiar, two-word rhythmic chant that erupted briskly somewhere in the crowd, every minute or so, and spread like thunder through space: "*Diretas já, diretas já*" (Elections now, elections now). If you have never taken part in a chorale formed by one million people, I recommend the experience. Nothing can prepare you for its penetrating sound, and nothing this side of the Milky Way will allow you to forget it. It is the sort of sound that carves memories for a lifetime.

Pressed by the ever-increasing flow of people, I climbed to the roof of a newsstand and, for the first time that night, gained a panoramic view of the entire citizenry that was conquering São Paulo's Anhangabaú Valley with its two-word song. For the long-vanished Tupi-Guarani, the native Indian tribe that inhabited that land before the Portuguese arrived in the sixteenth century, the stream that had run through the valley was known as the "river of the bad spirits." Not anymore. That night, the only river visible was a mighty Amazon of people. No bad spirit would have dared to exert itself in such a purposeful human ocean.

"What do we want?" part of the crowd spontaneously asked.

"*Diretas*" (elections), the rest of us answered.

"*Quando*" (when)? another group provoked.

"*Já, já, já!*" (now, now, now!) the whole crowd screamed back.

When that million-people choir began to sing the Brazilian anthem, not even the sky could hold its tears anymore. As the traditional São Paulo drizzle descended, I absorbed this resounding demonstration of what a population of individuals can do when they collaborate in harmony to achieve a common goal. Even though the message transmitted by the crowd (*Diretas já!*) was always the same, at any moment in time, a different combination of many voices was recruited to produce the crowd's roar. People weren't necessarily able to scream every time. Some were talking to their neighbors; others became temporarily hoarse, or were distracted while waving their flags; others simply dropped out of the chorale due to sheer emotion. Moreover, even as handfuls of people began to leave the rally later on, the crowd continued to thunder. For any observer, the loss of those few protesters did not make any difference at all—the overall potential population was so huge that the loss of a few people did not meaningfully alter the result.

Ultimately, the voices of those millions of Brazilians were heard. A few days later, I met with my mentor, Dr. César Timo-Iaria, to discuss a paper by David Hubel and Torsten Wiesel, who had shared the Nobel Prize in Physiology or Medicine in 1981 for their groundbreaking research on the visual cortex. Hubel and Wiesel had recorded the electrical activity of single neurons in the visual cortex, using the classical reductionist approach that was the norm in labs around the world at the time. I innocently asked Timo-Iaria why we did not do the same. His reply was as forceful as the roar that I had experienced as a member of the crowd in São Paulo: "We do not record from a single neuron, my son, for the same reason that the rally you attended a few days ago would be a disaster if, instead of one million people, only one person had showed up to protest," he said. "Do you think that anyone would pay attention to the plea of a single person screaming at a political rally? The same is true for the brain: it does not pay attention to the electrical screaming of a single noisy neuron. It needs many more of its cells singing together to decide what to do next."

Had I been more observant on that historic night in 1984, I may have understood that the dynamic social behavior of that thundering crowd had set before me most of the neurophysiological principles that I would obsessively investigate over the next quarter of a century. But instead of listening to a chorus of political protesters, I would be listening for the virtually unheard electrical symphonies created by large populations of neurons.

These neural ensembles would eventually provide the means for liberating a primate brain from its biological body. But in the mid-1980s, very few neuroscientists saw any reason to relinquish the reductionist experimental paradigm and its focus on single neurons. Perhaps this was because other scientific fields, including particle physics and molecular biology, had experienced extraordinary success with reductionism; in particle physics, for instance, the theory and ultimate discovery of smaller and smaller particles, such as quarks, proved to be a linchpin in the definition of the so-called standard model, which continues to be the basis of our understanding of the physical universe.

Roughly speaking, in mainstream twentieth-century neuroscience, the reductionist approach meant breaking the brain into individual regions that contained a high density of neurons, known as brain areas or nuclei, and then studying the individual neurons and their connections within and between each of these structures, one at a time and in great detail. It was hoped that once a large enough number of these neurons and their connections had been analyzed exhaustively, the accumulated information would explain how the central nervous system works as a whole. Allegiance to reductionism led most neuroscientists to dedicate their entire careers to describing the anatomical, physiological, biochemical, pharmacological, and molecular organization of individual neurons and their structural components. This painstaking and wonderful collective effort generated a tremendous wealth of data from which many outstanding discoveries and breakthroughs resulted. With the unfair benefit of hindsight, today one could argue that neuroscientists were trying to decipher the workings of the brain in the same way as an ecologist would attempt to study the physiology of a single tree at a time in order to understand the rain forest ecosystem, or an economist would monitor a single stock to predict the stock market, or a military dictator would try to arrest a single protester at a time to reduce the effectiveness of a million-strong Brazilian chorus chanting *diretas já* in 1984.

For an observer who benefits today from the century-old work of the true giants of brain research, it seems that what much of neuroscience still lacks is an experimental paradigm for dealing with the complexity of brain circuits. Today, systems formed by large numbers of interacting elements—things like a political movement, the global financial market, the Internet, the immune system, the planet's climate, and an ant colony—are known as complex systems whose most fundamental properties tend to emerge through the collective interactions of many elements. Typically, such complex systems do not reveal their intimate collective secrets when approached by the reductionist method. With its billions of interconnected neurons, whose interactions change from millisecond to millisecond, the human brain is an archetypal complex system.

Part of the neglect toward exploring the complexity of the brain could be justified by the tremendous experimental challenges involved in "listening" simultaneously to the electrical signals produced by large numbers of individual single neurons, distributed across multiple brain areas, in a behaving animal. For example, at the time Brazilian crowds were fighting for presidential elections, no one in the neuroscience community was sure what type of sensor could be implanted in the brains of animals so that many of these minute neuronal electrical signals could be sampled simultaneously for many days or weeks, while subjects performed a variety of behavioral tasks. Moreover, there was no electronic hardware or sufficiently powerful computer available that neurophysiologists could readily utilize to filter, amplify, display, and store the electrical activity generated by tens of individual neurons simultaneously. Neurophysiologists wondered, almost in despair, how they might choose which neurons to record from in each brain structure. Worst of all, nobody had any idea how to analyze the huge mountain of neurophysiological data that would be generated in case these technical bottlenecks could somehow be solved.

Paradoxically, few neuroscientists ever doubted that the astonishing feats accomplished by the human mind—from the production of artificial tools to the generation of self-awareness and consciousness—arise from the brain's huge number of neurons combined with their intricate pattern of massive parallel connectivity. But for decades, any attempt to tackle the technical hurdles to listen to brain symphonies was dismissed as a chimera, a high-tech experimental utopia that might only be realized through an effort on the scale of the Manhattan Project.

Essentially, all expressions of human nature ever produced, from a caveman's paintings to Mozart's symphonies and Einstein's view of the universe, emerge from the same source: the relentless dynamic toil of large populations of interconnected neurons. Not one of the numerous complex behaviors that are vital for the survival and prosperity of our species—or, for that matter, of our close and distant cousins, primates and mammals—can be enacted by the action of a single neuron, no matter how special this individual cell may be. Thus, despite the great deal we have learned about how single neurons look and function, and despite innumerous scientific achievements of brain research for the past century, the straightforward application of reductionism to brain research has proven to be insufficient and improper as a strategy to deliver the field's most cherished promise, a comprehensive theory of thinking.

All this means that the traditional and well-disseminated view of the brain, the one espoused in artful prose and beautiful illustrations in most of the neuroscience textbooks, can no longer stand. In much the same way that Einstein's theory of relativity revolutionized the classic view of the universe, the traditional single neuron–based theory of brain function needs to be categorically replaced by what amounts to a relativistic view of the mind.

The first step in proposing any new scientific theory is to define a proper level of analysis for investigating phenomena and testing one's hypothesis about them. This allows for *validating* or *falsifying* the proposed theory—the essence of the scientific method. I contend that the most appropriate approach to understanding thinking is to investigate the physiological principles that underlie the dynamic interactions of large distributed populations of neurons that define a brain circuit (see Fig. 1.1). Neurons transmit information to one another through long, projecting structures—their axons—which make discrete, noncontinuous contact (the synapse) with nerve cell bodies and their protoplasmic, treelike structures, called dendrites. In my view, while the single neuron is the basic anatomical and information processing-signaling unit of the brain, it is not capable of generating behaviors and, ultimately, thinking. Instead, the true functional unit of the central nervous system is a population of neurons, or

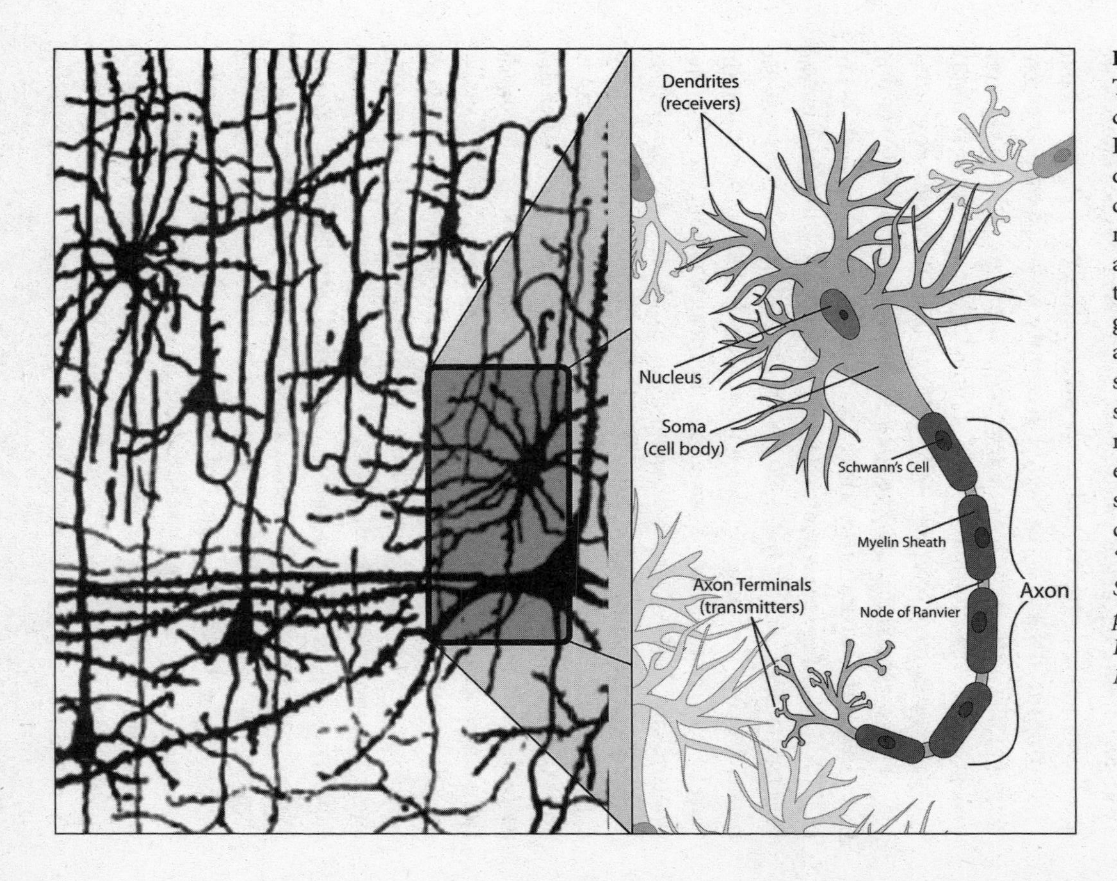

FIGURE 1.1
The architecture of a neural circuit. Reproduction of a Ramón y Cajal original drawing showing a neural circuit formed by many neurons. One single neuron and its cellular specializations are highlighted. In general, dendrites serve as the main neuronal specialization receiving synapses from other neurons. Axon terminals establish the neuron's synapses with other brain cells. *(Cajal's drawing from "Histology of the Nervous System" was reproduced with permission of the Cajal Legacy, Instituto Cajal [CSIC], Madrid, Spain.)*

neural ensembles or *cell assemblies*. Such a functional arrangement, in which populations of neurons rather than single cells account for the information needed for the generation of behaviors, is also commonly referred to as *distributed neuronal coding*.

Thinking with populations of neurons! Even two of humanity's most intimate possessions—a sense of self and a body image—are fluid, highly modifiable creations of the brain's mischievous deployment of electricity and a handful of chemicals. They both can change or be changed on less than a second's notice. And, as we will see, they do.

During the first half of the twentieth century, so-called single-neuron neurophysiologists argued, with seemingly incontrovertible evidence, that after sensory information was sampled from the external world through specialized receptors—the skin, retina, inner ear, nose, and tongue—it ascended through specific sensory nerve pathways that terminated in specific cortical areas. These areas were identified as the primary sites for processing sensory information in the cortex, with the somatosensory (tactile), visual, and auditory areas gaining particular prominence. During the same period, however, an American psychologist, Karl Lashley, emerged as the poster boy for the opposition: the distributionist camp. Lashley's main obsession was to identify the location in which the brain stores a memory, which he called the engram. In his experiments, he would surgically remove cortical tissue from various areas of the brains of rats, monkeys, and apes, both before and after the animals had been taught to perform particular behaviors, which ranged from simple tasks (learning how to identify a particular object visually and then jump or reach for it) to complex problem solving (learning how to navigate an elaborate maze). After an animal was trained, he measured the impact of the cortical lesions he had created on the animal's capacity to acquire or retain the behavioral skill, or habit, it had learned. With this experimental process, he aimed to understand how associations were built between sensory information and motor behavior.

According to Lashley, after animals had been trained in a simple task, much of the remaining cortex could be removed without affecting significantly the animal's behavioral performance—provided that some volume of the primary sensory cortex involved in the task was left intact. In fact, if just one-sixtieth of the primary visual cortex remained, the animal would retain a visual-motor habit it had learned. Faced with simple tasks, the brain was amazingly resilient in handling sensory information. In his classic article, "In Search of the Engram," Lashley summarized his results as the "principle of equipotentiality," whereby memory traces were distributed throughout the sensory area, not in a specific neuron or small group of neurons.

Yet, Lashley also had found that the brain was less able to recover from damage when faced with more complex behavioral tasks. An animal would begin to make task errors with a small number of lesions, and the amount of errors produced was proportional to the cortical mass removed surgically. Once 50 percent or more of the neocortex had been removed, the animal began to lose the learned habit altogether, requiring extensive retraining. Based on these findings, Lashley proposed a second principle of memory, the "mass action effect," which stated that "some physiological mode of organization or integrating activity is affected rather than specific associative bonds." Complex problem solving became "disordered" when parts of the cortex were taken out of commission.

Many neuroscientists criticized Lashley's conclusions. Even today, simply mentioning his name in a scientific talk invariably triggers all-knowing chortles of derision. Most of the scientific blowback was leveled at his experimental approach, particularly in trying to create brain lesions and then correlating them with too simplistic or too complex tasks. Still, Lashley was able to show that there was more going on in the primary sensory cortices than most neuroscientists were willing to acknowledge.

Usually, academic battles turn out to be so bloody because their stakes are often so miserably low. Not in this case. Defining the functional unit of the brain is a solemn endeavor. After all, this quest aims at pinpointing exactly which piece of organic matter decides, on your behalf, where your body starts and ends, what it feels like to be human, what deeper beliefs you hold, and how your children and the children of your children will one day remember who you were and what became of your legacy as a human being. Few human enterprises come close in relevance and drama as the ongoing search for the true reasons that make each of us feel so irrevocably different and unique, and yet so strikingly similar to our kin.

A simple analogy helps to clarify the distinction between the two competing views of brain function I am presenting here. Consider the role played by musicians in a symphony orchestra. If you had tickets to hear a concert and arrived to find out that just one bassoonist had showed up, you would be rather disappointed at the end of the night: no matter how proficient the musician had been, and how much effort he or she put into the performance, you would not be able to imagine the symphony's full score—not even if, instead of a bassoonist, the glorious violinist Anne-Sophie Mutter or the electrifying pianist Maria João Pires were onstage. You would only get an appreciation of the entire musical tapestry of the symphony if a significant number of musicians performed together, simultaneously. For the distributionists, when the brain creates a complex message or task using a large number of neurons, it is composing a type of symphony.

A neuronal concerto.

Coding a complex neuronal message or task into a large number of small, individual fragments or actions is similar to the work of an orchestra—each fragment helps to create the meaningful whole, like the million human voices singing "*diretas já*" that, with its sheer power, dethroned a dictator. This sort of distributed message strategy is often found in nature.

Distributed strategies are present in many aspects of our daily lives. For instance, the production of complex phenotypical traits—how our genetic makeup is expressed in, say, our physical appearance—often relies on the concurrent expression of many genes distributed across an array of chromosomes. Another natural distributed strategy involves multiprotein complexes, which operate within individual cells to perform a variety of functions ranging from DNA translation and repair to the release of chemicals, known as neurotransmitters, by neuronal synapses. Each protein is responsible for one specific subtask and many proteins may interact together to achieve a rather complex operation. For instance, different protein complexes, embedded in the width of the lipid membrane of a single neuron, form a variety of membrane ion channels. Each ion channel works like a tunnel through the membrane. When this tunnel opens, a particular ion (sodium, potassium, chloride, or calcium) can enter or exit the cell. Multiple ion channels cooperate to maintain or alter the electrical membrane potential of a single neuron. A single ion channel cannot regulate this process, just like a single neuron cannot produce a meaningful behavior. Instead, a population of diverse ion channels is needed for neuronal cell membranes to work properly.

Distributed strategies work at higher levels, too. For example, African lions usually hunt in packs, particularly when they want to capture large prey, such as a seemingly vulnerable elephant drinking at a water hole. This pack approach ensures that if one of the lions is killed by the elephant, the rest of the pack still have a chance to get that valuable elephant steak tartare by the end of the night. Conversely, some of the most preyed-upon species usually defend themselves from potential predators by gathering into dense groups when they roam in search of food for themselves. Thus, flocks of migratory birds crossing the thin air of the Himalayas, schools of fish navigating the glassy green shallows of the Caribbean, and swarms of capybaras, a South American rodent weighing more than one hundred pounds with menacing front teeth but little else to defend it, each rely on distributed strategies for protection against predators. By increasing the density of the pack of individuals traveling together, they divide the attention of their nemesis and significantly reduce the probability that a given individual will be caught. By doing so, the chance of perpetuating the group *as a whole* increases—a distributed strategy of risk management.

Does this approach to handling risk sound familiar? When financial managers advise you to diversify your portfolio, spreading your investments across a large number of companies representing multiple sectors of the economy, they are proposing exactly this sort of distributed strategy, without the capybara's menacing teeth. Even the most influential technology of our time, the Internet, relies on distributed computer grids to fulfill our apparently limitless thirst for information. No single computer controls the flow of bits and bytes across the whole system, and there's no single cable connecting your computer to Google's headquarters when you type in a request to find a Web page on a particular subject. Rather, huge numbers of interconnected machines very quickly route your Google search to one of the company's many computer servers in Mountain View, California. If one of these machines goes bonkers, no problem; the remaining computer network ensures that your query is not lost.

But why do distributed strategies work so well? Why, from proteins to packs of capybaras, does it make sense to rely on large and distributed populations of individual elements? To answer this fundamental question, let's return to the brain and examine the advantages of such a population-coding scheme for thinking.

By distributing thought across a large population of neurons, evolution has designed an insurance policy for the brain. In most cases, people do not lose an important brain function when a single neuron or a small chunk of brain tissue is damaged due to a localized trauma or a minor stroke. Indeed, because of distributed coding, a great deal of brain damage has to occur before patients exhibit any clinical signs and symptoms of neurological dysfunction. Conversely, imagine the risks you would incur if only one of the neurons in your entire brain was in charge of conveying a key aspect of your life, say, the name of your favorite Brazilian soccer team. Lose that neuron, and that information would be forever lost. Yet, throughout our adult lives, individual neurons continuously die without any major side effects. The fact that we almost never notice functional or behavioral effects, though these minuscule neuronal tragedies take place every day, speaks volumes in favor of distributed coding in the brain. Neuron populations are highly adaptive, or *plastic*; when damaged or dead neurons need to be bypassed, the remaining ones can self-reorganize, changing their physiological, morphological, and connectivity makeup when repetitively exposed to tasks and environments. As my friend Rodney Douglas, of the University of Zurich, recently noted, the brain truly works like an orchestra, but a unique one, in which the music it produces can almost instantaneously modify the configuration of its players and instruments and self-compose a whole new melody from this process.

Evolution may also have favored distributed population coding because it is far more efficient at delivering many complex messages than single-neuron coding. Let's take a simple example. Suppose that a single neuron can convey, or in neuroscience jargon, *represent*, two distinct messages by flipping between two frequencies of electrical firing, either very rapid firing or very slow firing. If just one neuron was devoted to detecting images in the visual field of an animal, the animal's brain would only be able to respond to two distinct images—firing at a rapid rate when one of the images was detected and at a slow rate for the other. Any other images would not be discernible by the single neuron at that same moment. Now suppose that one hundred different neurons were allocated to perform the same job. The number of distinct images that could be detected with the same two firing states would jump to 2^{100}.

In addition to this dramatic increase in computational power and memory, distributed coding in the brain relies on massive *parallel* information processing. Single neurons are capable of establishing an incredible number of connections by giving rise to axon processes that branch and reach many other different neurons simultaneously. This intricate mesh of neuronal connections can achieve wondrous things. For example, as part of my doctoral thesis, I created a simple computer program that could store, in a square matrix format, the direct connections shared between the pairs of brain areas and nuclei that form the circuit responsible for controlling cardiovascular functions. I then selected the most important forty brain structures that defined this circuit and identified which of the related forty nuclei were directly connected by a bundle of axons or nerves that uses only one synapse, called a monosynaptic pathway. In my computer program's forty-by-forty matrix, rows indicated the brain structures from which neurons gave rise to such monosynaptic pathways; the columns indicated the structures that received them. If structure number 4 had neurons that sent direct axonal projections to structure number 38, I noted "1" in the respective matrix position (intersection of row 4 and column 38). If neurons belonging to structure 38 reciprocated this connection and sent axons back to structure 4, another "1" was added to the intersection of row 38 and column 4. If there was no direct connection between a given pair (for instance, between nuclei 5 and 24), a "0" was added to the respective matrix position (see a reduced example in Fig. 1.2). Having gone through the trouble of building such a detailed matrix of direct, monosynaptic connections, I decided to ask a very simple question: given all the known pairwise connectivity of the circuit, how many neural pathways existed that could connect any pair in this circuit that did not have a direct monosynaptic connection? In other words, was there any way for information to flow between two unconnected pairs in the circuit? With that question in mind, I set a series of twenty IBM-PC XT microcomputers to run my program, hoping to get an answer. Each of these computers was supposed to seek potential multisynaptic pathways linking one of twenty distinct pairs of brain structures that did not share a direct monosynaptic pathway. At the end of this search, each computer would then print out the potential pathways in a list and a summary graph. I then headed out for a five-day holiday, to celebrate that most sacred of Brazilian religious events, Carnival.

Imagine my shock when, upon returning to the lab, I found that piles and piles of printouts had been generated by half of the computers. The programs running on those ten computers had identified thousands of potential multisynaptic pathways for connecting pairs of structures that did not talk directly to each other (Fig. 1.2). More surprising, of the other ten computers, some had not yet finished printing the potential pathways, while others had simply run out of paper. Even with just a handful of direct pairwise nuclei connections, there were hundreds of thousands or even millions of potential pathways for exchanging information between pairs of brain structures that did not share a monosynaptic connection.

By relying on large populations of interconnected neurons and massive parallel processing to encode information, advanced brains like ours become dynamic systems in which the whole becomes larger than the sum of its individual parts. That happens because the overall dynamic interactions of the network can generate complex global patterns of activity, known as *emergent properties*, which cannot be predicted from the outset by the linear sum of the individual features of individual elements. Such extreme nonlinear behavior enhances dramatically the physiological and behavioral outcomes that can emerge from the neural networks of the brain. Distributed networks formed by millions or even billions of neurons generate emergent properties such as brain oscillations, complex rhythmic firing patterns that underlie a variety of normal and pathological functions including certain states of sleep and epileptic seizures. Emergent brain properties also generate highly elaborate and complex brain functions, such as perception, motor control, dreaming, and a person's sense of self. Our very consciousness, arguably the most awesome

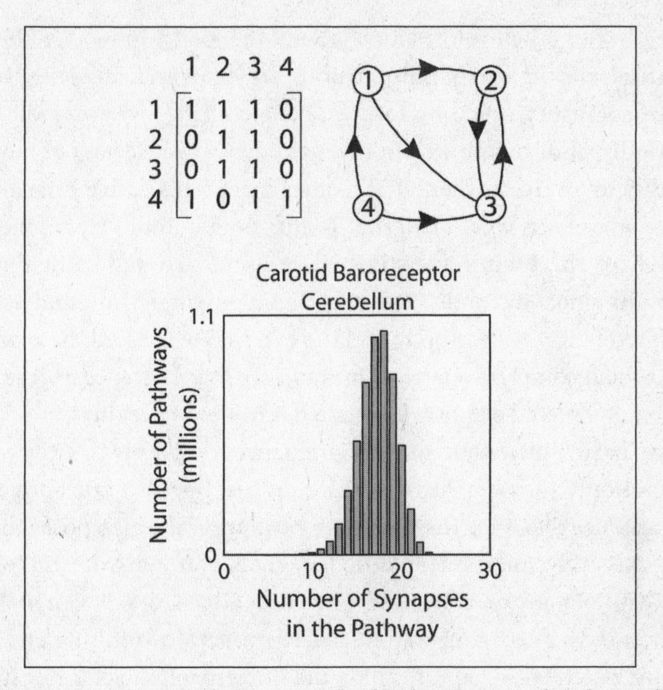

FIGURE 1.2 Use of graph theory to study the distribution of pathways linking pairs of neurons. On top, a square matrix is used to represent the direct, monosynaptic connectivity of a small brain circuit. In this matrix, 1 represents the existence of a direct connection between a pair of brain structures, while 0 depicts its absence. Next to the matrix, a graph is used to represent the circuit. Circles with numbers represent the structures and directional arrows represent the direct connectivity information contained in the square matrix. The histogram below depicts the total number of pathways linking two structures (carotid baroreceptor and cerebellum) that did not share a direct, monosynaptic connection. The *X* axis represents the number of synapses of the pathways and the *Y* axis depicts the number of pathways found. Notice that millions of pathways were found for this particular example. *(Courtesy of Dr. Miguel Nicolelis; redrawn by Dr. Nathan Fitzsimmons, Duke University.)*

endowment known to us, likely arises as an emergent property of a multitude of dynamically interacting neuronal circuits of the human brain.

But the new view of the brain I propose involves much more than a simple shift in emphasis from a single neuron to populations of connected brain cells. Up to now, most neurophysiological theories have consistently ignored the fact that highly elaborated brains do not sit tight and wait for things to happen. Instead, these brains take the initiative and actively gather information about the body in which they are embedded and its surrounding world, tirelessly and diligently sewing the cloth of reality, opinions, loves, and, I am afraid, even prejudices that we proudly, and sometimes blindly, wear every millisecond of our lives, blissfully unaware of where it all comes from. This active information-seeking maintains what I call the "brain's own point of view": the combination of the brain's accumulated evolutionary and individual life history, its global dynamic state at a given moment in time, and its internal representation of the body and the external world. All these components, which comprise our most intimate mental existence, merge into a comprehensive and exquisitely detailed rendition of reality.

The "brain's own point of view" influences decisively the way we perceive not only the complex world around us, but also our body image and our sense of being. The Cartesian assumption, which poses that our brains passively interpret or decode signals coming from the outside world, without any preconceived viewpoint attached to it, can no longer stand up to the experimental evidence. In fact, to fulfill its enormous scientific potential—from unveiling the intricate physiological principles that govern the operation of the human brain to developing brain-machine interfaces capable of both rehabilitating patients devastated by neurological diseases and greatly augmenting human reach—mainstream neuroscience must divest itself from its twentieth-century dogma and wholeheartedly embrace this new view.

In his masterpiece, *The Organization of Behavior*, published in 1949, the Canadian psychologist Donald O. Hebb promoted the concept that cell assemblies are the true functional unit of the nervous system. A student of Lashley's, Hebb also postulated that no "single nerve cell or pathway [is] essential to any habit or perception." He also pointed out that "electrophysiology of the central nervous system indicates . . . that the brain is continuously active, in all its parts. An afferent [incoming] excitation [from the body's periphery] must be superimposed on an already existent excitation [inside the brain]. It is therefore impossible that the consequence of a sensory event should often be uninfluenced by the pre-existent [brain] activity."

I propose that the brain's work results from the dynamic interplay of billions of individual neurons that create a continuum in which neuronal space and time seamlessly combine. In a fully behaving animal, as Hebb proposed, no incoming sensory stimulus is processed without first being compared against the brain's internal predispositions and expectations, arduously built through the collection of signals and memories from previous encounters with similar, and even not so similar, stimuli earlier in life. The diffuse electrical response evoked in the brain of a conscious subject when a novel message arrives from the periphery of the body seems to depend heavily on the internal state of the brain at that particular moment in time. Thus, while the constancy of the velocity of light determines why space and time have to be relativized in relation to the state of motion of a pair of observers in the universe, I contend that evolutionary and individual history, the fixed maximum amount of energy a brain can consume, and the maximum rate of neuronal firing offer the constraints that require an equivalent relativization of space and time within our heads.

Most information about the world and our body comes to the brain as a result of exploratory actions initiated by the brain itself. Perception is an *active* process that starts inside our heads and not at a peripheral site on the body with which the outside world happens to come in contact. Through a variety of exploratory behaviors, the brain continuously tests its own point of view against the new information it encounters. Even though we routinely experience the "feeling" on our fingertips of such tactile attributes as texture, shape, and temperature, in reality these sensations are skillful illusions crafted by the brain—"felt" during the split second in which our fingertips make contact with an object and collect and transmit the sensory data via the nerves back to the brain. If the feeling doesn't match the brain's expectations, it will correct the mismatch by creating a moment of surprise and discomfort, like the one that emerges when you reach into a package of bread and drop the slice when you find it is wet and slippery rather than dry and crumbly. The same process takes place during our elaborate, simultaneous visual, auditory, olfactory, and gustatory "experiences" of the world. All these indisputable human traits are borne by massive electrical brainstorms, which we usually refer to, more colloquially, as the act of thinking.

But could we push the definition of thinking any further? I believe so. I propose that the brain is actually the most awesome simulator to evolve in the known universe, at least as far as we can independently verify. Like a faithful and patient modeler of reality, the main business of our brains is to produce a large variety of behaviors that are vital for our existence as human beings. In essence, these physiological purposes boil down to:

(a) maintaining our bodies in working order through the global physiological process called homeostasis;
(b) building and storing very detailed models of the external world, of our lives, and of the continuous encounters between the two; and
(c) actively and continuously exploring the surrounding environment in search of new information to test and update these internal models. This includes learning from experience and predicting future events and their payoffs by generating potential expectations for their outcomes, costs, and benefits.

This short list covers most of the basic functions of the central nervous system.

By definition, a good simulation or model allows its user to continuously analyze and monitor all sorts of events to predict future outcomes. Neurophysiologists have spent a great deal of time investigating how the brain maintains the body's homeostasis, and in recent decades there's been an explosion of research into how the brain encodes sensory, motor, and cognitive information. But for the most part, because of the difficulty in studying these phenomena experimentally, they have avoided the highly complex behaviors that are encompassed in building and nurturing a model of the world, the pervasive and primordial human longing to create a detailed account, no matter how mystical and abstract, of how the universe was created, how humanity emerged, and why we have the gift of life in this otherwise mundane solar system. Often, these longings are left to the realm of religious inquiry. But these same complex behaviors also endow humans with an ardent curiosity, a key and unique trait of our species, which has led to the emergence of art as well as scientific thinking. Complex behaviors also encompass the elaborate social and courtship strategies employed by humans to achieve the evolutionary goal of transmitting genes to future generations, as well as our continuous attempts to imprint our ideas, dreams, beliefs, fears, and passions into the memories of our loved ones, friends, and other members of our species.

By now you may be thinking that the theoretical shift I am proposing is no big deal. Yet, this issue has played a central role in an ongoing theoretical scrimmage that has engulfed neuroscience during a two-hundred-year intellectual battle over the brain's soul. And as it happens, the notion of the brain as a model builder has found significant support outside the neuroscience community. In his classic book, *The Selfish Gene*, the British evolutionary biologist Richard Dawkins espouses the view that the brain, particularly that of humans, has evolved the enormously advantageous capacity of creating very elaborate simulations of reality. The physicist David Deutsch goes even further by proposing, in his book *The Fabric of Reality*, that all "we experience directly is a virtual-reality rendering, conveniently generated for us by our unconscious mind from sensory data plus complex inborn and acquired theories (i.e. programs) about how to interpret them."

In the first paragraph of his masterpiece book, *Cosmos*, Carl Sagan muses, "The Cosmos is all that is or ever was or ever will be. Our feeblest contemplations of the Cosmos stir us—there is a tingling in the spine, a catch in the voice, a faint sensation, as if a distant memory, of falling from a height. We know we are approaching the greatest of mysteries."

As far as we know, there is only one offspring of this awesome cosmos capable of deciphering its majestic language while producing a reel of luxurious sensations that our true parents, faraway deceased supernovas, never had the privilege to enjoy. While these progenitors burned away, unaware that their stardust would one day blow the breath of life on a small bluish planet revolving around an average star located in a remote corner of a distant galaxy, our brains endow us to lustily consume every bit of our conscious existence while silently carving in our minds the many intimate tales of a lifetime.

Thus, if ever there has been a scientific battle worth fighting for, it is the one in which neuroscientists have embroiled themselves for the past two centuries. And if you asked me to take a side, I would not hesitate a millisecond to say that, at the end of this intellectual brawl, as Brazilians proved twenty-five years ago, those siding with the inebriant plea generated by another huge crowd, one formed by billions of interconnected neurons, shall prevail.

2

BRAINSTORM CHASERS

Sir Edgar Douglas Adrian was keenly aware of the disputes that tormented the founding fathers of neuroscience. Installed at a lectern in the College of Saint Mary Magdalene at Oxford University in 1946, Adrian, himself a Cambridge man, set out to describe what in his opinion was the first major achievement in understanding the brain: an argument about the seat of "intelligence." He intoned that the brain "is a particular structure of nerve-cells and nerve-fibres, found in certain animals but not in all, and some animals which have no brain in the strict anatomical sense have none the less a complex behaviour, well adapted to the circumstances, a behaviour which we might well call intelligent. And within our own bodies there are many cells swimming freely in the blood and behaving as more or less independent living beings, avoiding what is bad for them and selecting what is good. Have they any claim to minds?" Adrian then related that more than two centuries earlier, at the end of the seventeenth century, philosophers at Cambridge and Oxford raged over the question of whether intelligence was situated in one part of the body—the brain—or throughout it. The Cambridge dons took the side of a singular site, while the Oxford dons looked farther afield.

To make his point, Adrian slyly quoted the opening stanzas of the poem "Alma: or, The Progress of the Mind," published in 1718 by Matthew

Prior, who had earlier been a lecturer in medicine at his very own Cambridge. After mocking the idea that the mind, in a "slap dash" manner beholden to ancient Aristotelian philosophy, "runs here and there like Hamlet's ghost" throughout the body, Prior describes the argument in favor of the brain:

> The Cambridge wits, you know, deny
> With ipse dixit to comply.
> They say (for in good truth they speak
> with small respect of that old Greek)
> That putting all his works together
> Tis three blue beans in one blue bladder.
> Alma they strenuously maintain
> Sits cock horse on her throne the brain
> And from that seat of thought dispenses
> Her sovereign pleasure to the senses

With this bravura performance of British wit, Adrian simultaneously teased his Oxford rivals while paying tribute to his Cambridge ancestors, the latter of whom had recognized that the brain was the sole culprit for the vicissitudes of the human mind. More than six decades later, I can almost imagine the circumscribed smirk quivering across the great man's face as he recited Prior's poem to the assembly.

Adrian was granted the latitude to make this academic jab. After all, he had been the very first neuroscientist to measure exactly how sensory information about the surrounding world and the body is coded by the peripheral nerves into the electrical signals that are the language of the brain—work that had won him a share of the Nobel Prize in 1932. An earlier researcher, Keith Lucas, had proposed that these electrical sparks, later named "action potentials," were all-or-none in nature. Adrian probed further into this idea and discovered that the intensity of the stimuli, whether tactile, olfactory, gustatory, or visual, related to the frequency of the action potentials transmitted in the peripheral nerves.

It was fitting then that in his lecture at Saint Mary Magdalene, Adrian also evoked the great dispute between two Italian scientists, the physician Luigi Galvani and the physicist Alessandro Volta, to revisit how electrophysiology was born of accident and soon became a significant field of scientific inquiry. Around 1783, Galvani had learned that by touching the leg muscle of a dead frog with two contacting strips, made of distinct types of metal, he could induce the muscle to contract. He interpreted this finding as proof that electricity was stored in the dead muscle fibers and that he had found the secret driving force of life. Volta was shocked by this naive conclusion and strongly, but respectfully, pointed out that electricity was most likely being created by the two distinct metal strips contacting the muscle and one another. Volta was a serious scientist, after all, so he knew that he would need to prove his point of view with evidence. According to Volta, the frog leg muscle was actually doubling as a conductor and a biological detector for the electrical current generated by Galvani's probe made of two distinct metals. Convinced of this interpretation, Volta went on to design the first electric battery, also known as a voltaic pile, by substituting brine-soaked paper for the frog's muscle as the conductive material filling the space between two metal plates made of zinc and silver.

As Adrian noted, it seemed that Volta was the clear winner of this electrical tempest, leaving poor Galvani with the wondrous prospect of being forever known as the senseless experimentalist who could not even correctly interpret his own results. Indeed, very few people today recall that Galvani should be credited for designing the first, rudimentary neuroprosthetic device, capable of artificially stimulating a nerve in a muscle—for this very same research. Fortunately for Galvani, other scientists soon obtained conclusive evidence that both living muscle and nervous tissue generated electrical currents. This discovery of animal electricity, though, really wasn't the major shock that Volta claimed it to be. In fact, the electrical currents were quite tiny, which explains why it proved so difficult for so long to measure them accurately.

Nature, it seems, does not write its riddles with just a few notes. More often than not, it tends to compose a symphony with a variety of tonal and rhythmic nuances that invariably sound strange and novel to our rather narrow perceptual skills. Scientists who, when faced with new evidence or fringe phenomena, resist modifying their pet theories usually find that the noise in the data, no matter how counterintuitive, is telling them something essential.

■ ■ ■

In a mind-boggling way, the fundamental debates in brain research not only resemble but are historically intertwined with the classic struggle among physicists to decide whether light should be considered to be formed by a wavelike phenomenon or particles, the latter theory favored by both Sir Isaac Newton and his gravitational nemesis, Albert Einstein. An extraordinary British physicist at Cambridge named Thomas Young—who was also an Egyptologist, linguist, physician, physiologist, and neuroscientist—played a key role in both of these scientific controversies.

Young's unprecedented scientific career tends to provoke the appropriate awe; one biography of him, written by Andrew Robinson, is titled *The Last Man Who Knew Everything: Thomas Young, the Anonymous Polymath Who Proved Newton Wrong, Explained How We See, Cured the Sick, and Deciphered the Rosetta Stone, Among Other Feats of Genius.* One of those feats involved his ingenious and now classic double-slit experiment, which is now known simply as Young's experiment. By flashing light through a thin plate that contained two parallel vertical slits separated by a short distance, Young observed that an alternating pattern of bright and dark bands appeared on a screen placed behind the slits. Because this pattern resembled the "interference pattern" observed when two waves of water, produced by two stones thrown into a lake at the same time, collide, Young proposed that light was in fact a wave. Many physicists, like the genial Richard Feynman, credit Young's double-slit experiment as the founding event of quantum mechanics.

Incredibly, just a year after this revolutionary experiment, Young began formulating what became his theory of distributed neural coding, called the trichromatic theory of color vision. Not too shabby for the eldest of ten children of a Quaker family roosted in Milverton, Somerset.

I was initially introduced to Young's work as a result of my friendship with Robert Erickson, whom I met shortly after I arrived at Duke, where he was a senior professor in the Department of Psychology. A well-known gustatory physiologist, Erickson was an ardent supporter of the distributionist notion that brains relied on neural populations to encode information. He was also one of the few people alive to have traced the origin of the debate between localizationists and distributionists in neuroscience to the disputes between Thomas Young and the phrenologist Franz Gall.

By all accounts, Gall was himself an accomplished anatomist. Just a couple of years before Young published his trichromatic theory in the *Philosophical Transactions of the Royal Society*, Gall began to popularize his purportedly clinical method, then called "cranioscopy," for identifying fundamental character traits and mental skills through careful analysis of a person's skull. Gall argued that certain areas of the cerebral cortex would grow disproportionately in people endowed with certain artistic skills, mental abilities, and aberrant behaviors. This localized differential growth of the cortex, according to Gall, would impinge upon the actual form of the skull, allowing a skillful examiner, like himself, to palpate a person's head and announce the unique aptitudes and shortcomings of his or her brain, even whether it would make the person a gifted writer—or a cold-blooded murderer. Gall divided the brain into twenty-seven "organs" (translated as skull bumps), nineteen of which he said were shared by all animals, up to and including humans. In addition to the organs dedicated to basic emotions such as the instinct of reproduction, the love of one's offspring, pride, arrogance, vanity, and ambition, there were organs specified to dictate a person's religion, poetic talent, and firmness of purpose and perseverance. According to Gall's schema, people endowed with an unusually accurate memory exhibited protruding eyeballs.

During his lifetime, most of the medical and scientific community strongly disagreed with Gall's outlandish conclusions. That did not stop Gall and his disciples from disseminating his ideas in lectures all over Europe, particularly the notion that mental functions are spatially localized in specialized modules in the cortex. Yet, as Robert Erickson always mentioned in his papers, there is no escaping from history: localizationist neuroscience is the legacy of Gall, just as distributionist neuroscience is the legacy of Young.

Although he never mentioned it to me during our years together on the Duke faculty, I later discovered that Erickson carried the heritage of another scientific dynasty. Erickson had studied under Carl Pfaffmann, whose groundbreaking work on the gustatory nerves in cats provided evidence that, even at the level of peripheral nerves, information could only be coded by the concurrent activation of many broadly tuned nervous fibers. As Erickson later recounted in one of his review articles, Pfaffmann asserted that "in such a [gustatory] system, sensory quality does not depend simply on the 'all or nothing' activation of some particular fiber groups alone, but on the pattern of other fibers active."

Pfaffmann's lab was located in the physiology department at Cambridge University. In a small footnote to the article, Erickson shares the delightful story of how his mentor got his start in gustatory research. At Cambridge, Pfaffmann collaborated on research with Lord Adrian. By then, Adrian had basically claimed for himself the investigation of almost all of the peripheral sensory nerves, as well as their central projections. His empire covered the visual, auditory, olfactory, and somatosensory (touch) systems. The only system Adrian did not grab for himself was taste. As Erickson put it, "This he assigned to Pfaffmann." Erickson was clearly gratified that he belonged to such a distinguished scientific family tree. His pride was evident when, prior to quoting the original formulation of Young's trichromatic theory, he stated that this "hypothesis for the encoding of color is succinctly put in what are arguably the two most powerful sentences in the history of neuroscience."

Essentially, without any other source of information but that provided by Young's brilliant logic, the trichromatic theory of color vision predicted the existence of three distinct types of receptors for color vision in the human eye. Here is the first of the two sentences lauded by Erickson, with his clarifications of the scientific jargon noted in parentheses, in which Young defined the trichromatic theory in 1802:

> Now, as it is almost impossible to conceive each sensitive point of the retina to contain an infinite number of particles (receptors), each capable of vibrating in perfect unison (responding) with every possible undulation (wavelength), it becomes necessary to suppose the number limited; for instance to the three principal colors, red, yellow, and blue, of which the undulations are related in magnitude nearly as the numbers 8, 7 and 6; and that each of the particles is capable of being put in motion less or more forcibly, by undulations differing less or more from a perfect unison; for instance, the undulations of green light, being nearly in the ratio of 6½, will affect equally the particles in unison with yellow

> and blue, and produce the same effect as a light composed of those two species; and each filament of the nerve may consist of three portions, one for each principal color.

Five years later, Young would go further, arguing that "the different proportions in which they (the sensations) may be combined, afford a variety of tints beyond all calculations." It took some time, but late in the twentieth century experimentalists finally demonstrated the existence of Young's retinal color receptors—the three types of retinal cones.

In his extremely rich book, *Origins of Neuroscience*, the historian and neuroscientist Stanley Finger details how Young's trichromatic theory was rescued from anonymity by the physician and physicist Hermann von Helmholtz, who provided both the data and a more mathematical formulation to validate it fully. Finger also argues that Young's theory inspired Johannes Müller to develop his theory of specific nerve energies. This theory postulates that the perception of different sensations emerges as a direct result of the stimulation of particular receptors and nerves. But as much as I enjoyed Finger's book, I cannot agree with him on this last assessment. If anything, Young's theory suggests the opposite—that is, that a particular sensation, in this case color vision, depends on a pattern of activation across many distinct nerve fibers.

This idea is usually understood more readily by examining a graphic display of Young's model for color-coding in the retina (Fig. 2.1). This graph depicts the bell-shaped response curves of the three retinal receptors postulated by Young. Although each of the receptors responds maximally to the presence of one of three main colors (blue, green, or red), it also has the capacity to respond, albeit at lower and graded magnitudes, to the presence of other colors. That is the definition of a broadly tuned receptor, or neuron: a biological "detector" capable of responding submaximally to the presence of a broad range of values of a given physical entity, such as light, pressure, sound, or chemical concentration, and maximally to a particular value of this range.

An important detail to keep in mind is that the bell-shaped curves that distinguish the response profiles of the three retinal receptors overlap considerably in the dimension of wavelength (color). This means that a given color stimulus is likely to trigger a response, albeit a different

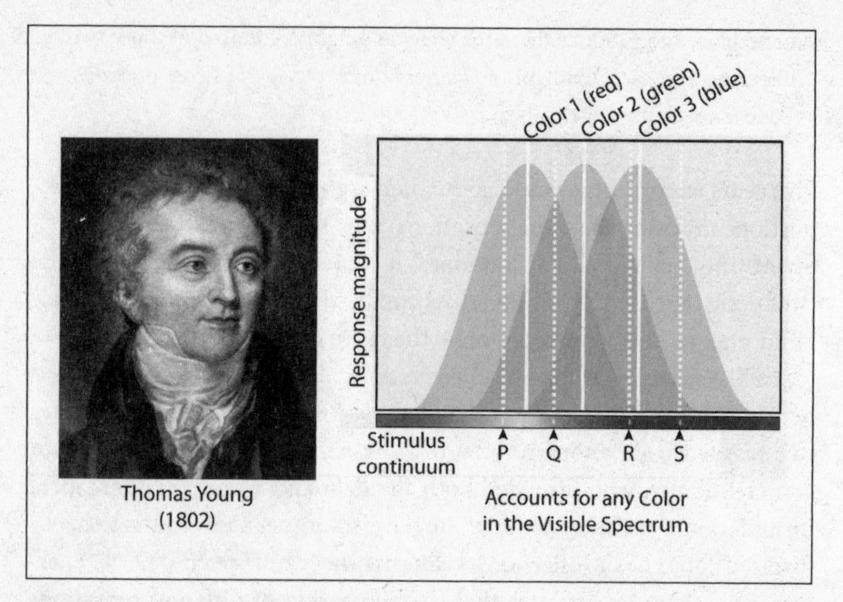

FIGURE 2.1 Thomas Young's trichromatic theory of color. A portrait of Thomas Young and a graphic representation of his theory on the right. Notice that any color stimulus (P, Q, R, or S in the *X* axis of the plot on the bottom left) can be represented by the graded response of three distinct "color receptors," which responds maximally to red, green, and blue respectively, but can also respond submaximally to different colors. *(Young's portrait was reproduced with permission of the National Portrait Gallery, London. The figure was originally published in M.A.L. Nicolelis, "Brain-Machine Interfaces to Restore Motor Function and Probe Neural Circuits."* Nature Reviews Neuroscience *4 [2003]: 417–22.)*

one, from each of the three receptors. This graph also shows how Young's three retinal receptors would collaborate to indicate the presence of a large set of distinct colors. For each color stimulus delivered to the retina, there is a particular distributed population pattern produced by the sum of the distinct electrical firings generated by each of the three retinal receptors. For example, in the case of a distinct color "P," the unique retinal response pattern is defined by an almost maximal signal from receptor 1, a 20 percent magnitude signal from receptor 2, and a nil response from receptor 3. Conversely, the color "Q" induces an almost maximal signal from receptors 1 and 2, and a 20 percent magnitude signal from receptor 3. By using just three "broadly" tuned receptors, the retina acquires the power to represent the presence of an incomprehensibly large set of colors. As we saw in chapter 1, this is one of the most remarkable advantages of distributed neural coding: the stunning capability of representing a huge number of messages that far exceeds, by many orders of magnitude, in fact, the number of elements employed by the neuronal population. And without any of today's high-tech instruments, Thomas Young imagined this capability. Just by thinking!

We now know that Young's broadly tuned neurons exist throughout the primate brain. This fact was not known in the nineteenth century, and a great battle for the "functional soul" of the brain was waged by neuroscientists. In 1861, however, the localizationists scored a mighty blow with the publication of a clinical study by the French physician Paul Broca. He reported the case of a patient who had experienced a profound loss of fluent speech—he was only able to utter the nonsense word "tan," no matter what he was trying to say—who was also severely paralyzed down the right side of his body. The man died soon after Broca examined him, so the patient's brain was recovered and dissected. To his astonishment, Broca found extensive lesions in the middle, convex section of the brain's left frontal lobe. Here was hard evidence that mental functions were discretely localized in distinct parts of the brain. Broca quickly tried to dissociate his finding from phrenology, stating that this "speech" center was not in the same place as the skull bump postulated many decades earlier by Gall and his progeny. Yet, as Finger shrewdly notes, both language centers were located in the frontal lobe, and that was enough. There was no escaping Gall's bulging-eyed ghost.

Broca's discovery caused a commotion in the medical community, and many neurologists became converts to the ideology of a brain comprised of specialized functional areas. Nine years later, two German scientists, Eduard Hitzig and Gustav Fritsch, delivered what appeared to be the true coup de grâce. By applying mild electrical currents sequentially to different regions of the frontal cortex in dogs, Hitzig and Fritsch elicited obvious muscle contractions in distinct parts of the animals' bodies. The researchers also demonstrated that surgically removing a particular region in one hemisphere of the cortex produced a noticeable, though not complete, deficit in the strength and maneuverability of the dogs' forepaws on the opposite side of the body. Hitzig and Fritsch used this data to sketch out a complete motor map of a dog's body in a well-defined part of the frontal lobe that became known as the motor cortex. Such topographic representations of an animal's body are also known as somatotopic maps. A century later, we know that our brains contain several of these maps, not just in the frontal lobe, but also in several areas within the parietal lobes and in many subcortical structures.

Spectacular as these findings were, they were soon overshadowed by the growing influence of a new breed of experts armed with microscopes and brain tissue stained by chemical reactions. These histologists would stage their decisive showdown with the last remaining general of the distributionist opposition in the nineteenth century, during the ceremonies marking the award of the 1906 Nobel Prize in Physiology or Medicine.

Like any year, the twelve months of 1906 witnessed their fair share of tragedies, triumphs, and memorable human achievements. In April, San Francisco was shaken by the horrendous earthquake that embroiled the city and killed more than three thousand of its residents. In August, another earthquake reduced to rubble the city of Valparaiso, the coastal resort in Chile, where three thousand people died. In Italy, Mount Vesuvius erupted. Lava, incandescent rock, and ashes spread over Pompeii and Naples, killing hundreds and dislodging thousands of people.

The night before the San Francisco earthquake hit, the great Italian tenor Enrico Caruso sang the role of José in the opera *Carmen* at the Tivoli Opera House. Awakened by the enormous jolt of the quake, Caruso ran down the stairs of the Palace Hotel to seek refuge in the street. Legend has it that, holding a picture autographed by President Roosevelt as his only form of identification, Caruso managed to escape the burning city by gaining passage on a boat to New York City. In November, Caruso was brought to face a New York City judge and accused of an indecent act allegedly committed in the monkey house of the New York City Central Park Zoo. Mrs. Hannah Graham alleged that Caruso had pinched her bottom—unceremoniously and uninvitedly. In his defense, the singer, making sure he spoke in a way that would preserve his voice for an upcoming staging of *La Bohème* at the Metropolitan Opera, claimed that most likely Mrs. Graham was pinched by a monkey. The judge did not buy his story. Caruso was fined ten dollars and released.

Caruso's best American friend, President Theodore Roosevelt, also had a busy year. After becoming the first U.S. president to travel abroad to visit the "ditch," his favorite nickname for the Panama Canal, in December, to his own astonishment, he was informed that he had been awarded the Nobel Peace Prize for his role in mediating a truce agreement in the war between Russia and Japan. To his delight, he was now part of a very special and select group of individuals and an official guest to a ceremony that, by all accounts, changed the future of brain research.

On the characteristically chilly but scientifically red-hot Swedish night of December 10, 1906, the great hall of the Swedish Royal Academy of Music in Stockholm was filled by the nation's royal family, members of parliament, distinguished scientists, and—as one of the laureates of the night wrote in his memoir—"many elegant ladies." They, along with members of the late Alfred Nobel's family, waited solemnly for His Majesty the King of Sweden to formally award that year's Nobel Prizes. Likely, few of the people assembled would have been unaware of the tension between the two men who would share the prize given in medicine that evening—the first time the Nobel Committee had not selected a single individual to receive the honor.

Count Karl Axel Mörner, the rector of the Karolinska Institute, which had been charged with selecting the prize recipients, announced the winners. He pronounced, "This year's Nobel Prize for Physiology or Medicine is presented for work accomplished in the field of anatomy. It has been awarded to Professors Camillo Golgi of Pavia and Santiago Ramón y Cajal of Madrid in recognition of their work on the anatomy of the nervous system." As the count continued to recite his graciously worded presentation, he reminded his listeners—including the men honored that night—about how much of the brain's activity remained a mystery. After a brief foray into describing the complex anatomy of the nervous system, he returned to the prizewinners, who, he said, had given birth to an entirely new branch of medicine.

When it finally came time to discuss the specific research that had brought Golgi and Ramón y Cajal to Stockholm, a sneaky international diplomacy prevailed. Addressing Golgi in Italian, Count Mörner said, "Professor Golgi. The Staff of Professors of the Karolinska Institute, deeming you to be the pioneer of modern research in the nervous system, wishes therefore, in the annual award of the Nobel Prize for Medicine, to pay tribute to your outstanding ability and in such fashion to assist in perpetuating a name which by your discoveries you have written indelibly into the history of anatomy." Mörner then turned to Cajal and switched to Spanish. "Señor Don Santiago Ramón y Cajal. By reason of your numerous discoveries and learned investigations, you have given the study of the nervous system the form that it has taken at the present day, and by means of the rich material which your work has given to the study of neuroanatomy, you have laid down a firm foundation for the further development of this branch of science," he said. "The Staff of Professors of the Karolinska Institute is pleased to honor such meritorious work by conferring upon you this year's Nobel Prize." With those historic words, neuroscience was baptized.

Born in Petilla de Aragón, Santiago Ramón y Cajal was obsessive and autocratic, a stubborn genius who single-handedly launched brain research into its modern era by demonstrating unequivocally that brains, like other organs, are made of collections of individual cells. A fanatical microscopist, he combined exquisite technical skills with a panache for drawing and creative insights.

Few would believe, if told, that this man receiving the Nobel award from the hands of King Oscar II that night had published his first paper on the central nervous system only eighteen years earlier, when he was an anonymous professor at the University of Valencia. Early in his career, Cajal had not been able to write in German, the language of the leading anatomists of his time. So he founded a scientific magazine, *Revista Trimestral de Histología Normal y Patológica*, in which he could report his findings in Spanish. It helped that he continued to finance and edit the publication throughout its existence—it made for a very straightforward editorial review. In 1896, he repeated the experiment and created the first classic scientific journal of brain research, *Revista Trimestral Micrográfica*. The debut issue contained six papers authored by Don Santiago himself. Years later, German anatomists decided to learn Spanish just to be able to read Cajal's original masterpieces.

To this day, Cajal remains one of the most frequently cited authors in neuroscience. His ingenuity and resilience as an experimentalist were first on display when he adapted a staining method, called *la reazione nera* (the black reaction), for his revolutionary investigation of the organization of brain tissue. Cajal ceaselessly tinkered with the black reaction, hardening blocks of brain tissue in a compound of either potassium or ammonia dichromate crystals, and then moving the tissue to a solution of silver nitrate, where the tissue structures would slowly "blacken" against the translucent yellow of the firm block. Cajal then sectioned the blocks into thin sections, which could be viewed under a microscope. The patient histologist could then easily identify the splendorous brain-cell bodies, dendrites, and axons scattered across the tissue. Cajal tried using the technique on embryonic and newborn brain tissue as well as adult specimens. It took years for him to find the right combination of solution baths, slice thicknesses, and brain tissues. His most stupendous results came from the brains of birds, reptiles, and small mammals that he mostly caught himself and then prepared in his wife's kitchen. There were likely very few chickens pecking around the Cajals' backyard.

Having perfected the black reaction, he next developed a new illustration method, which consisted of drawing on a single sheet of paper every single individual cell he could observe as he shifted the focal plane of each slice of tissue he placed under his beloved Zeiss microscope (Fig. 2.2). These comprehensive microscopic images of brain tissue were stunning and unprecedented. His precise and pioneering images of brain circuits gave an inkling that neuroscientists would soon be able to disclose the intimate secrets and long-lost tales of the human mind.

Year by year, Cajal unveiled a series of unique discoveries about the morphology of the brain's cells, each embellished by his inventive illustrations and interpretations. These coalesced into his famous law of dynamic polarization, which stated that nerve cells were functionally polarized, that is, they were formed by a recipient region—represented by their dendrites—and a transmitting component—the axon. Using this concept, Cajal predicted that an electrical impulse would be received by the dendrites of a cell and, after passing through the cell body, be transmitted via the axons to other cells. Although there is no evidence that he ever witnessed a physiological recording of an electrical potential traveling across an axon or firing up from a dendrite or a cell body, three decades later electrophysiological recordings would prove him to be right. The story told in Madrid, where Cajal later worked, was that on

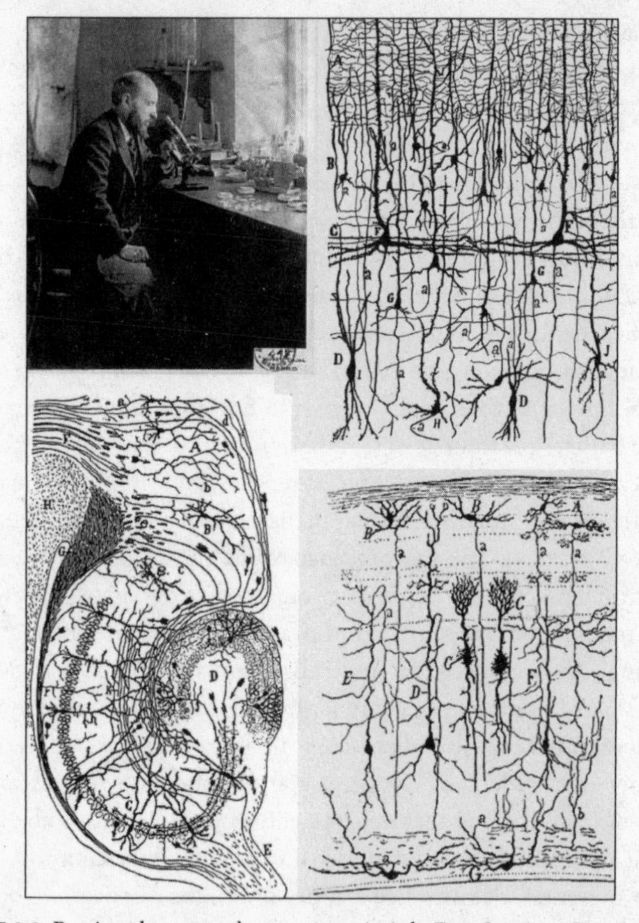

FIGURE 2.2 Putting the neuron's supremacy in ink. Don Santiago Ramón y Cajal at his favorite place, in front of a microscope, and a few of his masterpiece drawings of different portions of the central nervous system. *(Cajal's three drawings from "Histology of the Nervous System" and the photo were reproduced with permission of the Cajal Legacy, Instituto Cajal [CSIC], Madrid, Spain.)*

the day God decided to create the brain, He got very excited and decided to phone Professor Cajal to describe His ideas on how the brain should work. After attentively listening to God's plan, Don Santiago simply said: "Not bad—but, please, come down to Madrid and let me show Thee some slides so that Thou might learn how the brain Thou are about to create will actually work."

The crown jewel of Cajal's career was the enunciation of a series of laws that defined what became known as the neuron doctrine. According to his theory, the brain is formed by large numbers of individual cells that communicate with one another by discrete contacts. To Cajal's chagrin, though, he had not himself come up with the name "neuron" to designate this critical unit of the nervous system's anatomy. In 1891, a German anatomist, Wilhelm von Waldeyer-Hartz, had done so, in a review article that received a great deal of attention. But the name was coined, and Cajal had to endure it.

Given the dimension of Cajal's achievements, you may be wondering why in heaven the Nobel committee decided that he must share the prize for medicine with Camillo Golgi. In total opposition to Cajal's conclusions, Golgi ardently supported the competing theory of the brain, the "reticular theory" originally proposed by the German anatomist Joseph von Gerlach. The reticular theory postulated that the brain was formed not by individual cells but by a continuous, vast mesh of brain tissue. While Gerlach favored the idea that fused dendrites were the main component of this reticulum, Golgi believed fused axons or widespread neural networks—so-called nerve nets—primarily served to organize the brain's tissue, making Golgi one of the few neuroscientists of his day to abhor the idea that mental functions were localized in discrete regions of the cortex. He was out of step with the scientific times. But as it happened, Golgi had invented the crucial *reazione nera*, back in 1873. The Nobel committee could not ignore him. It was an ironic destiny, in the tradition of Galvani: an Italian, seemingly the last descendant of Thomas Young, had devised a fabulous new method but seemed to have badly misunderstood his own data. Indeed, despite their shared accolade, the practice of *la reazione nera*, a love for using their wives' kitchen as a lab, and a devotion to Zeiss microscopes were the only things on Earth that united Cajal and Golgi.

By the time of the Nobel presentation on December 10, Cajal and the neuron doctrine had long dominated the newborn field of neuroscience; the experimental evidence was overwhelmingly in the Spaniard's favor. Yet Golgi resisted. In fact, on December 11 he delivered a defiant Nobel lecture, titled "The Neuron Doctrine—Theory and Facts," in which he argued that the doctrine would soon be in decline. He then dissected the theory, point by point, occasionally raising his own ideas as an alternate approach. Midway through his presentation, he openly mocked those who took the neuron doctrine for granted: "I shall therefore confine myself to saying that, while I admire the brilliancy of the doctrine which is a worthy product of the high intellect of my illustrious Spanish colleague, I cannot agree with him on some points of an anatomical nature which are, for the theory, of fundamental importance."

The next day, a very upset Cajal took his place at the front of the Swedish lecture hall, armed with his irrefutably elegant pictures of distinct neurons and their delicate processes. He began by invoking the "tradition" of the Nobel lectures as a time for a scientist to present his own results. But Cajal was not above drawing blood. "We mourn this scientist who, in the last years of a life so well filled, suffered the injustice of seeing a phalanx of young experimenters treat his most elegant and original discoveries as errors," he said in closing.

For most observers, Cajal won the day. Throughout the twentieth century, localizationists would proceed to divide the cerebral cortex into visual, auditory, tactile, motor, olfactory, and gustatory centers. These original sites were then subdivided into specialized regions for color, motion detection, face recognition, and other complex functions. Soon, individual neurons were being labeled as visual neurons, mirror neurons, face neurons, touch neurons, even grandmother neurons.

Few areas were left uncharted, but how a whole brain worked remained a deep and obscure mystery. After dividing and subdividing the brain into its minute units, neuroscientists still lacked a way to explain how those units came together to produce the seamless perceptual experiences that define human life. Ironically, like his countryman Galvani, it now appears Golgi may have accurately seen the big picture, even though he could not identify it in the details of his black reaction slides. Moreover, in recent decades, scientists have even found that several areas of the brain, including a structure called the inferior olive, which is involved in motor control, and some classes of neurons, such as the inhibitory interneurons of the cortex and the mitral cells in the olfactory bulb, do form continuous networks. These networks are linked by cytoplasmic bridges known as gap junctions, similar to Golgi's version of the reticular theory. In a very unexpected way, though, Golgi did exact his revenge, albeit almost in silence: he helped coin the term *nerve network* and the general concept of a brain that thinks through the collective work of vast distributed neuronal circuits. The "Golgi neural network" that was so ridiculed in 1906, as it turned out, inspired generations of distributionists, including Lashley, Pfaffmann, Hebb, and Erickson, to resist and endure.

Galvani's and Golgi's stories remind me of what a famous Brazilian soccer coach once said: "These Italians can win a fight in very surprising ways." Just ask the millions of Brazilian fans—who watched in despair as the Italian striker Paolo Rossi scored three straight goals and kicked their team out of the 1982 World Cup—whether they still have nightmares about that game.

I certainly do.

3

THE SIMULATED BODY

Following the crowning triumph of the neuron doctrine during the 1906 Nobel ceremony, neuroscience witnessed an unstoppable and, for the most part, unmatched rise of the localization view of brain function. This impetus was particularly intense among those who focused on unraveling the organization of the cortex, the highly convoluted block of tissue that comprises the outer layer of the cerebral hemispheres. The opening years of the twentieth century saw the rise of cytoarchitectonics, which relied on a variety of staining techniques, including the Nissl method of dying the negatively charged RNA found in organelles within cells, to study the distribution and clustering of neurons.

Cytoarchitectonics was launched into prime time, at least in part, by the Russian histologist Vladimir Betz's discovery in 1874 that the motor cortex, the cortical region that Hitzig and Fritsch had tagged as the source of body movements, contained a peculiar horizontal layer packed with large pyramid-shaped neurons. Known since then as Betz cells, these pyramidal neurons give off very long axons that bundle together and descend all the way to the spinal cord, forming the corticospinal tract, one of the bulkiest and most important neural pathways. The corticospinal tract carries voluntary motor signals, generated in the motor cortex, to pools of interneurons, the cells that give rise to axons that make connections locally, and lower motor neurons located in the brain stem and the spinal cord. The axons of the lower motor neurons in the brain stem terminate in our facial muscles, while those in the spinal cord project to muscles throughout the rest of the body. When lower motor neurons fire electrical bursts, our body muscles readily oblige and contract. By transmitting detailed motor instructions to these lower motor neurons, the corticospinal tract exerts executive control on the generation of concrete movements. The corticospinal tract allows our inner voluntary motor intentions to be communicated to the surrounding world.

Cytoarchitectonic studies carried out at the end of the nineteenth century had determined that the cortex could be divided into six layers of neurons, each lamina stacked atop another. These cortical layers are numbered, from the outer to the inner layer, using the roman numerals I through VI. By measuring the thickness and number of layers per cortical area, the density and distribution of different cortical cell types across these layers, and other parameters, several histologists had proposed schemes for dividing the cortex into distinct areas or fields by the early twentieth century. Among these pioneers, the German neurologist Korbinian Brodmann introduced, in a series of papers published between 1903 and 1914, a comprehensive cytoarchitectonic classification, based on Nissl staining, suggesting that the mammalian cortex comprised fifty-two cortical fields (Fig. 3.1). Though in one of his initial studies Brodmann reported data from a single lemur's cortex, in his classic 1909 paper he described and documented with illustrations data obtained from multiple animal species. Based on these findings, he identified forty-nine distinct cortical areas in the human brain.

In Brodmann's classification scheme, each of the cortical fields was identified by a number. In some cases, the particular distribution of neurons in a given cortical layer provided the main feature for allocating the field's number and function. For example, Brodmann noted the prominent occurrence of Betz cells in layer V, in which his own area 4 was located. Area 4, in his view, contained the primary motor cortex. Similarly, layer IV appeared to be densely packed with cortical neurons that

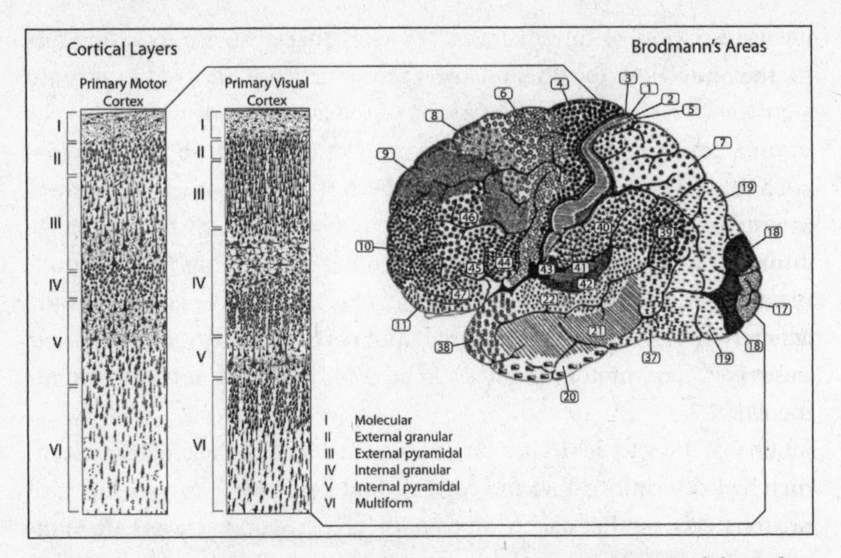

FIGURE 3.1 Korbinian Brodmann's cortical cytoarchitectonic map and the six layers of the cortex. A side view of the human brain with the original numeral designations of cortical areas created by Brodmann is on the right. A comparison of the six layers of a section through the primary motor (M1) and visual (V1) cortices is shown on the left. Cytoarchitectonically speaking, the M1 is characterized by the presence of large pyramidal-looking neurons (Betz cells) in layer V, while V1 exhibits a very dense cluster of neurons in the bottom part of layer IV and upper part of layer VI. *(Cajal's two drawings from "Histology of the Nervous System" were reproduced with permission of the Cajal Legacy, Instituto Cajal [CSIC], Madrid, Spain. Brodmann's Areas, originally published in 1910, are in the public domain.)*

were the main final target of the major sensory pathways (tactile, visual, and auditory), which carry information from the body periphery to the cortex. Brodmann used the existence of this densely packed layer IV in different cortical regions to identify the primary somatosensory (areas 3, 1, and 2), visual (area 17), and auditory (areas 41 and 42) cortices. The anatomophysiological correlations identified by Brodmann have stood the test of time. Yet, cytoarchitectonics suffered from its own obsession with more and more intricate subdivisions of the cortex. Indeed, at the same time that Brodmann was publishing his studies, his own teachers, Cécile and Oskar Vogt, proposed an alternative scheme that included more than two hundred distinct cortical areas. Even as cytoarchitects moved to other features and techniques—staining myelin fibers, for instance—

they were not able to provide a definitive coherent functional guide to how the brain works in a behaving beast.

Sir Charles Sherrington is considered by many to be the father of modern systems neuroscience. During the first two decades of the twentieth century, he and a series of students and collaborators at Oxford University employed a physiological approach to study the cortex. At the time, that meant relying on electrical stimulation of cortical regions while measuring the animal's behavior. Using this method, Sherrington and his colleagues demonstrated that the cortex of the primate frontal lobe contained a complete "motor map" of the animal's body. These studies, which were summarized in an eighty-seven-page paper published in the *Quarterly Journal of Experimental Physiology* in 1917, included experiments with twenty-two chimpanzees, three gorillas, and three orangutans. Sherrington had found the primary motor cortex of great apes in the precentral gyrus, a cortical area located in front of the central sulcus, which separates the frontal and parietal lobes.

The true impact of these studies was only felt when one of Sherrington's students, the American neurosurgeon Wilder Penfield, shared some unusual observations taken during neurosurgical procedures on patients suffering from epileptic seizures. After studying with Sherrington, Penfield had interned under the legendary American neurosurgeon Harvey Cushing at Yale, where he honed his skills in the operating room. Penfield then moved to Montreal to work at McGill University, where he founded and directed the Montreal Neurological Institute. Over a period of nineteen years, he collected data from more than four hundred craniotomies, in which, under local anesthesia, a piece of a patient's skull is removed, allowing the cortex to be exposed. Since manipulation or electrical stimulation of the cortex does not generate any pain, Penfield's patients remained conscious during the operation and could report what they felt when he stimulated different spots in their cortex in his attempt to locate the source of their epileptic seizures. During the procedure, Penfield and his collaborators, including the Canadian psychologist Donald Hebb, were able to map the type of tactile sensations elicited by electrical stimulation of cortical regions located in front of (precentral) and behind (postcentral) the central sulcus. Penfield found that 75 percent of the spots that provoked tactile sensations were located in the postcentral cortex, which contained the primary somatosensory cortex according to Brodmann; the remaining 25 percent were located in the precentral gyrus, where the primary motor cortex resides. More surprising, the sensations elicited by stimulation of the precentral cortex continued even in the few cases that required that the surgeons remove the postcentral gyrus in order to quell seizures. He also reported that stimulating the postcentral gyrus created body movements, even when the precentral gyrus was removed. Penfield believed this provided evidence that the sensations reported when the motor cortex was stimulated were not generated by collateral electrical activity or creeping nerve fibers from the primary somatosensory cortex. The primary motor and somatosensory cortices appeared to share their functions. While each area exhibited a clear functional bias, the two sides of the central sulcus contributed to similar sensorimotor behaviors.

Such an arrangement suggests the notion that cortical areas could exhibit a significant degree of functional specialization (in this case, either more motor responses elicited from the motor cortex or more tactile sensations from the somatosensory cortex) while still contributing to other brain functions or behaviors. In this context, a given cortical area, like the primary motor cortex, would normally have a higher probability of participating in the executive generation of motor behaviors while contributing in a secondary way to the genesis of tactile sensations. Conversely, under normal conditions, the primary somatosensory cortex would have a much higher probability of being involved with the generation of tactile sensations than motor responses. Yet, contrary to the dominant dogma sustained by the defenders of a strict segregation of functions in the cortex, the chance that neurons in the primary motor or somatosensory cortices could be recruited for these other tasks would not be zero.

Penfield proceeded to test what he had learned from his epileptic patients. He reconstructed the sequence of body sensations reported, slowly moving the spot at which he electrically stimulated the brain from the medial to the lateral portion of the postcentral cortex, immediately behind the central sulcus. He discovered that as he shifted the stimulation, the location of the tactile sensations progressively moved, too, starting at the toes, then the foot, then the leg, hip, trunk, neck, head, shoulder, arm, elbow, forearm, wrist, hand, each of the fingers, the face, the lips, the intraoral cavity, and finally the throat and the intra-abdominal cavity. When plotted across a cross section of the cortex, this sequence provided a topographic map of the human body, which became known as the sensory "homunculus." Although Penfield wrote the study, the depiction of the homunculus was drawn by Mrs. H. P. Cantlie. It took her two tries to satisfy the neurosurgeon. But in the end, she pulled off what became one of the most reproduced illustrations in the annals of medical literature (Fig. 3.2, left side).

The homunculus that so pleased Penfield does not resemble anyone you might see walking down the street or sitting at your dinner table. Instead, Mrs. Cantlie's homunculus appears severely, eerily distorted. This distortion is the result of a developmental process called cortical magnification that overrepresents the areas of the body with the highest density of mechanoreceptors, a series of highly adapted peripheral nerve endings responsible for translating tactile stimuli into electrical potentials, the language of the brain. So the homunculus bulges at the fingers, hands, and face, most especially the area around the mouth and the tongue. Other body areas, including the chest and the trunk, seem squeezed—as though they are on a sensory diet—despite occupying larger proportions of our skin. Because the fingers, hands, and face contain so many mechanoreceptors, they are our most refined tactile organs, the ones we normally use to create a tactile image of our surrounding world. That's also why it's so difficult to tell what an object is when it's rubbed across the skin of the back.

The phenomenon of cortical magnification is not a privilege of the human species. In every mammal examined in the past seventy years, it has been documented extensively. In rats, the "rattunculus" involves a large representation of the array of facial whiskers, with the forepaws supersized compared to the rear ones (Fig. 3.2, right side). In the case of the platypus, the semiaquatic, egg-laying Australian mammal, it is the animal's beak that is overmagnified in the somatosensory cortex's body map.

These somatotopic representations are also not unique to the cortex.

FIGURE 3.2 The homunculus meets the rattunculus. The drawing depicts an impossible meeting between the reconstruction of the cortical "homunculus," the distorted representation of a human's body in the primary somatosensory cortex, based on Wilder Penfield's studies, and a cortical "rattunculus," the equivalent distorted representations of the rat's body, in the rodent primary somatosensory cortex. Notice the overrepresentation of the lips and hands in the homunculus and the facial whiskers, snout, and forepaws in the rattunculus. The cheese is Swiss. *(Illustrated by Dr. Nathan Fitzsimmons, Duke University.)*

Each of the subcortical relays of the bundle of axons that form the somatosensory nerve pathways carrying tactile information from the body periphery and proprioceptive feedback from muscles and tendons to the central nervous system contain such maps. For this reason, it seemed that these topographic maps were a fundamental physiological tool employed by the brain to shape tactile perception—but then a paradox emerges. There is no doubt that our most extraordinary tactile experience is a sense of inhabiting a body that we recognize as our own. From just a few months of age, humans are able to distinguish their own bodies from other objects and other people. And throughout our lives, we experience and interact with the world from our body's first-person perspective. Yet, none of our routine tactile experiences, not even this most private and meaningful one, resembles Mrs. Cantlie's homunculus. The drawing feels eerie because it doesn't match our sense of self.

So how in heaven is the body image we actually experience defined, if not by the toil of the homunculus imprinted in the brain? To begin to answer this question, we have to enter the mental landscape of "experienced" phenomena—near-death encounters and phantom limbs—that cannot be explained by orthodox, localizationist neuroscience.

One morning, late in my final year at the University of São Paulo medical school, one of my good friends, a young vascular surgeon working at the Hospital das Clinicas, its main teaching hospital, invited me to visit the orthopedics ward. His invitation was unusual, to say the least.

"Today we will talk to a ghost," he said in a solemn tone. "Do not get frightened. Try to stay calm. The patient has not accepted what has happened yet, and he is very shaken."

Of course, I had never met a ghost in my life, though my Italian great-grandmother had repeatedly assured me that phantasms floated invisibly around us and did not particularly like children who refused to go to bed before the end of the Wednesday night soccer broadcast. I decided to check Dona Ada's theory empirically.

Despite many previous calls at the hospital emergency room, which on some nights resembled a rumbling war zone, I was not prepared for my visit to the always austere orthopedic institute. As we entered a small, isolated infirmary, we were received by the tired gaze of a stocky middle-aged woman who had just risen from her chair and was sobbing. The woman's round and reddish face was covered with deep lines; together with her leathery hands, they betrayed a life of sorrow and hardship. Sitting beside her in a reclined bed was a boy of about twelve years, his face dripping with sweat and contorted in an expression of horror. The child's body, which I now watched closely, writhed from excruciating pain.

"It really hurts, Doctor; it burns without a break. It seems as if something is crushing my leg," he said. I felt a lump in my throat, slowly strangling me.

"Where does it hurt?" I ventured.

He did not hesitate: "In my left foot, my calf, the whole leg, everywhere below my knee!"

Moved by an uncontrollable reflex, and a sickening feeling of revolt that a child had been lying in pain in a hospital bed in the middle of the day, I started to lift the sheets that covered his sweat-soaked body. My indignation was immediately replaced by disorientation when I noticed that half of the boy's left leg was missing. As my colleague later told me, the boy's leg had been amputated below the knee after he had been run over by a car.

Outside the ward, my surgeon friend tried to calm me. "It was not him speaking," he said. "It was his phantom limb."

At that time I did not realize that at least 90 percent of amputees—millions worldwide—experience these phantom limb sensations: the uncanny feeling that a missing body part is still present and attached to their bodies. In some cases, the part moves; in others, it is locked in place. Usually, such ghostly appendages are often defined by a diffuse tingling sensation that extends throughout the amputated limb and effectively reconstructs it. These phantoms are often very painful and terrifyingly vivid. In some cases, they persist for years.

The phenomenon of phantom limbs has been reported for centuries. During the Middle Ages, European folklore glorified the restoration of sensation in amputated limbs in soldiers. In one classic account of mysterious cures in the port of Aegea, in the Roman province of Siria, during the fourth century, patients who had lost their arms and legs supposedly sensed the angelic presence of their missing limbs thanks to the "miraculous" works of twin brothers who were later canonized as saints by the Catholic Church. According to the Church's documentation, Saints Cosmas and Damian had "restored" the sensation of a missing leg by transplanting the leg of a dead person into the stump. Legend had it that any amputee who evoked the brothers' names would, once again, feel their missing limbs.

In the sixteenth century, phantom limbs moved from the realm of religion to that of medicine. The French military surgeon Ambroise Paré, whose improved surgical techniques boosted the survival rate for amputees, noted many cases of phantom limbs among soldiers returning from European battlefields. Although he believed his patients, Paré feared that people would think he had lost his mind while attending them—which may explain why he published his findings in French and not in Latin, the scientific language of his day, and why his observations were neglected for more than three centuries.

That neglect may provide some insight into the particular heroism of the British admiral Horatio Nelson, whose legacy includes a remarkable self-description of a phantom limb. During the Battle of Santa Cruz de Tenerife, in 1797, soon after coming off a small boat and landing on shore, Nelson was shot in the right arm by a Spanish musket ball. The injury was severe. Most of his arm was amputated.

Eight years later, on the eve of the Battle of Trafalgar, Lord Nelson foresaw the victory of the British armada over the combined French and Spanish navies. In a letter he drafted to his queen, he revealed that he had received a divine presage: a vivid sensation that he could still hold high the sword with which he had sworn to defend the Crown of England—with the limb he had lost in Tenerife. The next morning, armed with his ghost sword (and more than two thousand cannons, I may add), Nelson defeated Napoleon's forces. Later that day, he died, the victim of a shot that this time was lethal.

The modern clinical investigation of phantom limbs, however, had to wait for a bloodier battle. A few days after the Battle of Gettysburg, the American neurologist Silas Weir Mitchell documented case after case of phantom limbs, many among Confederate amputees who felt compelled to reenact their participation in Pickett's charge, an uphill assault on the Union army and the deadliest skirmish of the battle. Hopelessly lying in their medical barrack beds, the amputees experienced the sensation of endless throbbing, almost like the pain of running past the point of no return. It was Mitchell who coined the term "phantom limb."

Interviews with thousands of amputees have been written up since the Civil War. Their cases suggest that severe limb pain before the amputation, due to a severe fracture, deep ulcer, burn, or gangrene, is a major risk factor for later developing phantom pain. More than 70 percent of patients find their phantom limbs painful immediately after surgery; in up to 60 percent, the throbbing persists for years. Phantom limbs sometimes perform phantom movements. Recent amputees may even wake up screaming that their nonexistent legs are trying to walk off the bed and run away on their own. In one-third of afflicted people, the absent limb becomes completely paralyzed, often agonizingly so—for instance, embedded in an ice cube, permanently twisted in a spiral, or tortuously bent against the back.

Researchers now know that phantom sensations can occur in any excised body part, not only the arms or legs; people who have lost their breasts, teeth, genitals, and even internal organs can experience them. Women with hysterectomies have reported illusory menstrual pain and uterine contractions, similar to those felt during labor. Curiously, male transvestites who have undergone sex reassignment surgery do not experience a phantom penis, which suggests that, to their brains, these men already live in a woman's body.

Despite intense investigation over the past century, neuroscientists have yet to pin down the origins of phantom limbs. An early hypothesis, put forth by the British neuroscientist Patrick Wall, who held a professorship at the Massachusetts Institute of Technology (MIT), proposed that the phantom limb phenomenon originated as a result of spurious activity generated by severed nerve fibers in the scarred region of the stump. These severed fibers formed nodules, or neuromas, which were thought to send erroneous signals through the spinal cord to the brain. Based on Wall's hypothesis, neurosurgeons began to design treatments aimed at removing this peripheral source for the misinterpreted signals. But when they cut the sensory nerves leading to the spinal cord, severed the nerves in the spinal cord itself, or even removed the parts of the brain that received the sensory neuronal tracts, the phantoms persisted. A patient's pain might vanish temporarily, but it always returned—with a vengeance. As these clinical observations accumulated, many neuroscientists rejected the notion that neuromas, or any other abnormality at the level of peripheral nerves, could explain the richness of symptoms of the phantom limb syndrome.

The main voice of opposition came from the great Canadian psychologist Ronald Melzack, who had studied under Donald Hebb. In 1965, Melzack and Patrick Wall were working together at MIT when they introduced a daring proposition, which became known as the gate control theory of pain. According to this theory, the pain sensation associated with a peripheral noxious stimulus, that is, a stimulus that generates some kind of body injury, can be modulated, or "gated out," at the level of the spinal cord. This happens when there is concurrent activity in other peripheral nerve fibers, such as those that carry light touch information, or even in nerves that descend from the cortex and other higher brain centers to the spinal cord but are not themselves associated with signaling pain. Melzack postulated that the central brain structures played a fundamental role in the control of pain sensation. His hunch was dramatically proven a few years later when significant analgesia was produced by electrically stimulating the periaqueductal gray matter, a small region buried deep in the brain whose neurons extend axons to the very spinal cord regions where peripheral "pain fibers" converge. Next, researchers found that this analgesic effect was mediated by endogenous opiates produced by the brain cells—endorphins.

This chain of discoveries, initially triggered by the gate control theory of pain, revolutionized pain research by unequivocally demonstrating that the sensation of pain is an internal construct of the brain. The brain is capable of modulating a noxious stimulus arriving from the periphery, according to its will as the ultimate modeler of reality (and its sorrows). This shifted the reference point for understanding pain from the peripheral pain receptors and nerves to the brain's own point of view. Neurobiology could now begin to explain why soldiers filled with a belief in a true and legitimate moral cause (e.g., ridding the world of the Nazis) continue to fight for their countries after suffering devastating, painful wounds; why marathoners continue to run, mile after mile, despite throbbing foot injuries; and why, as Melzack demonstrated experimentally, Italian mothers tend to scream much more than Irish ones during regular baby deliveries.

Following their work on the gate control theory, in the 1980s Melzack and his colleagues developed an alternative explanation for the phantom limb phenomenon: the elaborate illusions experienced by amputees emerged not from peripheral neuromas but from widely distributed neuronal activity within the patient's brain. Dedicated "pain fibers" and "pain pathways," as the localizationists had called them, simply did not exist. Instead, pain—and the entire realm of sensations and emotions associated with it—exemplifies the way in which the products of our brain's intricate neuronal circuits are conceived, interconnected, informed, and delivered to our consciousness. Pain can strike suddenly, when you first notice blood dripping, even though the cut was sliced several moments or minutes earlier. And it can grow into anguish, and sometimes out of control, as it matures into a lingering memory.

Melzack's explanation of phantom limbs challenged the classic dogma of perception. He proposed that in addition to detecting sensory signals from the body, the brain also generates a pattern of activity, or "neural signature," that defines the body's image or schema at any given moment of our lives. He suggested that this internal brain representation, extending well beyond the reach of the homunculus that Penfield had identified in the motor and somatosensory cortices, endows each of us with a sense of our own body's configuration and borders and establishes the definition of our very sense of self. According to Melzack, the brain's image of the body and its limits would persist even after the removal of a body part, creating the anomalous yet vivid sensation of a phantom limb.

The dynamic sculpturing of the neural signature, according to his new theory, falls to a large network of neurons that Melzack has named the "neuromatrix." The neuromatrix includes the somatosensory cortex, located at the surface of the brain on the top of the head, and its associated regions of the parietal lobe. In addition, it encompasses multiple neural pathways, including one that conveys tactile information from the periphery of the body to the thalamus, a sensory relay station deep in the brain whose neurons send information to the somatosensory cortex, and another that traverses the brain's limbic system, a group of buried brain structures that governs emotions such as those associated with phantom limbs.

Damage to part of the neuromatrix can result in a loss of ownership of part or all of one's body. For instance, extensive injuries to the right parietal lobe caused by brain trauma, a tumor, or a stroke can lead to a complex neurological condition, known as left hemibody neglect syndrome, in which patients become indifferent to the entire left side of their bodies and, in most cases, the environment surrounding that side. Patients with the condition have been known to fail to put on the left sleeve of a shirt or a left shoe. When asked about their behavior, they typically deny that the left arm or leg is theirs—it belongs to someone else's body.

The clinical symptoms of this syndrome are usually transient, but they can cause commotion. Take an incident described to me by a NASA astronaut who visited my lab at Duke several years ago. During the initial orbit of his first space mission, he said, the pilot of the space shuttle started complaining to his colleagues. "Stop poking your hand on my left control panel!" When the pilot was told that nobody was poking their hand in his panel and that the hand in question was his own left hand, he shrugged and replied, "The hand on the left panel is certainly not mine." A few hours later, to the relief of the rest of the crew (and mission control in Houston), the pilot suddenly reported: "Just relax, guys. I have found my missing left hand on the control panel!"

Melzack argues that the basic structure of the neuromatrix may be already present at birth, its blueprint defined by genetic instructions. This congenital network would explain why, as Melzack reported in 1997, phantom arms and legs appear in at least one-fifth of children born without these body parts, and in half of the children whose limbs were amputated at a very young age. These remarkable findings suggested that the human brain is capable of generating a well-defined model of the subject's self even in the absence of somatic sensory signals derived from the physical body.

For more than half a century, since Penfield's neurosurgical observations, the existence of the homunculus map of the body's surface had been accepted. Neuroscientists generally believed that this somatotopic map, like other topographic representations that had been found in the primary visual and auditory cortices, was only malleable during a brief period of early postnatal development known as the critical period. After that point, the consensus was that the brain's topographic map "crystallized," remaining stable throughout the rest of the person's life. This notion was based on evidence provided by Nobel Prize winners David Hubel and Torsten Wiesel regarding the segregation of ocular dominance columns, clusters of neurons of the primary visual cortex that carry signals from either the left or right eye. Based on this work, cortical maps were believed to be incapable of exhibiting "plastic" functional reorganization during adulthood.

That picture started to change in 1983 when two American neuroscientists, Jon Kaas of Vanderbilt University (Fig. 3.3, left panel) and Michael Merzenich of the University of California, San Francisco, announced that traumatic amputation of the middle finger of an adult monkey led to a remarkable functional reorganization of the somatotopic map in the primary somatosensory cortex. Instead of remaining silent after the amputation, after a few weeks or months the cortical neurons that had represented the finger started to respond to any tactile stimulus delivered to an adjacent region of the hand, such as the index and ring fingers (Fig. 3.3, right panel). All of a sudden, old monkey neurons could learn new tricks. Few people realized that more than a decade earlier, Patrick Wall, who had originated the gate control theory of pain, and his students had published a small study in the journal *Nature* claiming to have induced plasticity in the somatosensory thalamus, the main subcortical relay of the neural pathways that carry tactile information from the skin to the cortex, in adult rats.

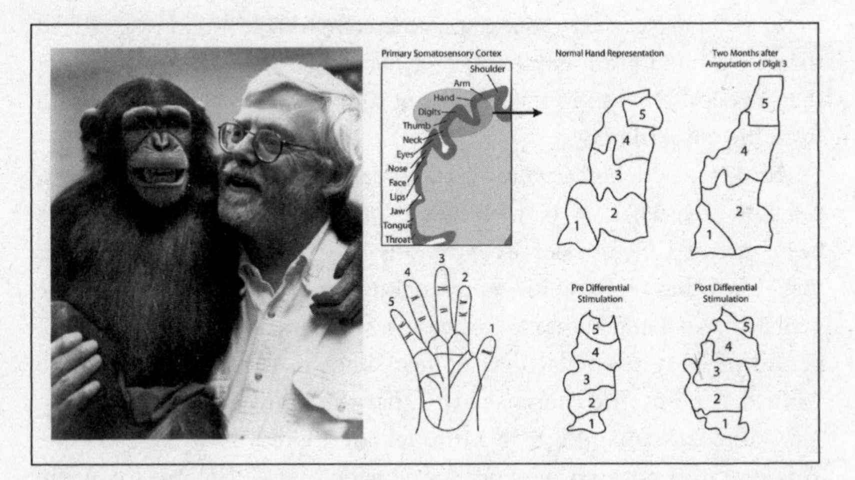

FIGURE 3.3 Revolutionary plasticity experiment in owl monkeys by Jon Kaas and Michael Merzenich. On the left, Jon Kaas appears with a close collaborator. On the right, the top shelf describes how, after a traumatic amputation of the third finger, the territory representing this finger in the primary somatosensory cortex of an owl monkey does not remain silent. Instead, the region that was occupied by finger 3 is now invaded by enlarged representations of fingers 2 and 4. The lower shelf shows that enlargement of the representations of fingers 1–4 can be obtained by repetitive, selective stimulations of these digits in detriment of finger 5. To see the effect, just compare the map on the bottom middle (before selective stimulation) with the cortical map on the bottom right (after stimulation). *(Adapted from, M. M. Merzenich, J. H. Kaas, J. Wall, R. I. Nelson, M. Sur, and D. Felleman, "Topographic Reorganization of Somatosensory Cortical Areas 3B and 1 in Adult Monkeys Following Restricted Deafferentation,"* Neuroscience *8, no. 1 [1983]: 33–55, with permission from Elsevier.)*

The findings of Kaas and Merzenich triggered a true revolution in the field. Clearly, the mammalian brain had evolved to be plastic. Yet, some of the later proponents of cortical plasticity subsequently resisted the notion that subcortical structures were capable of functional reorganization, too. So it caused a stir when in 1993, my postdoctoral research adviser, John Chapin, now at SUNY Brooklyn, and I showed that the plastic reorganization process started *immediately* after a small dose of local anesthetic was injected subcutaneously to block neural activity from a small area of skin (and thus more easily—and less traumatically—mimic the effects induced by a finger amputation). Our studies showed that the reorganization process occurred at the subcortical level, in structures such as the thalamus. Soon afterward, Tim Pons, a neuroscientist then at the National Institutes of Health (NIH), conducted studies in monkeys whose entire arm—not just a finger—had been deafferented many years prior. Deafferentation involves severing the connection between all sensory afferents (nerves) and the spinal cord. Pons reported that long-term deafferentation prompted a widespread reorganization, in which the neurons previously assigned to the hand switched to react to signals from the face, which is represented next to the arm in the brain's map. He and his colleagues also noted that the reorganization process occurred in the thalamus and brain-stem relays of the somatosensory system.

What do all these observations in monkeys have to do with a potential explanation for the occurrence of phantom limbs in amputees? The connection became evident when V. S. Ramachandran, a physician and neuroscientist at the University of California, San Diego, documented the occurrence of plastic reorganization of the topographic body map present in the somatosensory cortex of patients whose arms had been amputated. Using an imaging technique called magnetoencephalography (MEG), which measures the magnetic fields produced by electrical activity in the brain, in the early 1990s Ramachandran and his colleagues showed that tactile stimulation delivered to regions of these patients' faces activated what was supposed to be the hand area of their cortical body map. As Ramachandran explains in his enlightening book, *Phantoms in the Brain*, when he touched particular spots on the amputees' faces, the patients instantly claimed that they felt sensations in their phantom hands. Furthermore, Ramachandran's team found that tactile stimulation of specific points on the face elicited sensations at specific points on the phantom hand. The type of sensation—whether hot, cold, rubbing, or massaging—was the same in both locations. The brains of these patients connected their existing face to their phantom hands. The link between adult brain plasticity and phantom limb pain was corroborated in a 1995 study in which neuroscientist Herta Flor at the University of Heidelberg in Germany and her colleagues employed MEG to detect the degree of cortical reorganization in thirteen amputees. There was a strong correlation between the amount of functional cortical restructuring and the magnitude of phantom pain.

Spurred by this evidence, Ramachandran and his collaborators developed a very simple but ingenious treatment for phantom limb syndrome based on two main theoretical pillars: first, that the adult brain's body maps are malleable, and second, that the brain's internal workings, and not the feed-forward flow of tactile signals from the peripheral nervous system, dictate the sculpturing and maintenance of the perception of body identity and uniqueness. Their therapy involved building a "mirror box" that patients could use to practice calming a phantom arm. A vertical mirror was inserted into a cardboard box from which the top had been removed. The amputees were then instructed to insert their intact arm in the front side of the box so that the arm's reflection in the mirror overlaid the perceived location of the phantom limb. This created a visual illusion that the phantom arm had been resurrected, almost like the miraculous effect credited to Saints Cosmas and Damian. In this case, though, the effect was experimentally recorded—and had long-lasting results. When patients moved their existing arms, they felt the "phantom" arm obey these same motor commands. Six of the patients who used the mirror box said they could feel as well as see their phantom arm moving, generating the impression that both arms could now be moved. Four of the patients used this newfound ability to relax and open a clenched phantom hand, providing relief from painful spasms. In one patient, a routine of ten minutes of practice per day with the mirror caused his phantom arm and elbow to "disappear" completely within three weeks. And when the limb vanished so did the pain. The visual illusion apparently corrected the tactile one, suggesting that the activity of the central visual circuits could modify the activity of Melzack's neuromatrix.

Almost a decade later, psychologist Eric Brodie of Glasgow Caledonian University and his colleagues reported hints of success in a test of a mirror box modified for phantom legs. Forty-one lower-limb amputees watched a reflection of movements of their intact leg in a mirror as they tried to move their phantom leg. Another thirty-nine amputees tried to move both their phantom and real legs without the mirror. Both efforts, which involved ten different movements each repeated ten times, diminished phantom limb sensations, including pain. Although the mirror did not enhance this effect, it did produce significantly more phantom limb movements and more vivid awareness of the phantom leg than did the exercise without the mirror. Prolonged mirror treatment might be effective in fighting phantom pain, Brodie proposes, because it reverses the plastic reorganization of the brain.

Researchers are now trying to ameliorate phantom limb pain with immersive, three-dimensional computer simulations—so-called virtual reality (VR)—that can produce illusions similar to those created by a mirror. The technology can display a patient's entire body, including the phantom limb, and enable a patient to perform complex movements of the fingers, toes, hands, feet, arms, and legs that are not possible with mirror therapy alone. In a preliminary study conducted in 2007, psychologist Craig Murray at the University of Manchester and his colleagues exposed two upper-limb amputees and one lower-limb amputee to a simulation that transposed the users' existing limb movements onto those of a virtual limb, which overlaid their phantom limb in the virtual environment. All three amputees, who each participated in two to five VR sessions, reported feeling sensations in the phantom. In each case, phantom pain decreased during at least one of the sessions, suggesting that VR therapy might offer pain relief as well.

The clinical evidence obtained from these patients underscores that our body image, that inexpugnable refuge of our carefully groomed individuality and mental uniqueness, emerges graciously as a dynamic by-product of the collective electrical activity of brain circuits to remain malleable and responsive to events occurring within, on, and beyond the physical boundaries of our mortal skin. Just like any good and sensible modeler of reality, the brain has endowed us with what feels like a true and concrete physical instantiation of the self, a simulated body.

But if our body image is just a simulation, how does the brain create and maintain such a convincing illusion over the course of a whole life? How easily can such an internal neural model be changed, and how far can the limits of our self reach?

Recent experiments have started to address these key questions and, to the surprise of many in the neuroscience community, the emerging answers are stunning. The conclusion from more than two decades of experiments is that the brain creates a sense of body ownership through a highly adaptive, multimodal process, which can, through straightforward manipulations of visual, tactile, and body position (also known as proprioception) sensory feedback, induce each of us, in a matter of seconds, to accept another whole new body as being the home of our conscious existence.

Take, for instance, the so-called rubber hand illusion, first demonstrated by cognitive neuroscientist Jonathan Cohen, now at Princeton University (Fig. 3.4). Subjects are asked to sit in a chair and rest their left arms close to the left edge of a small table placed in front of them. An opaque screen is then positioned in front of the person in order to hide the resting left arm from his or her sight. Then, a full-size dummy rubber arm and hand are placed on the same table, but a bit closer to the subject. Care is taken to place the dummy arm and hand so that they occupy a position where the subject could plausibly imagine that they might be his or her own arm—which remains out of sight, behind the screen. The subject is then told to fix his or her gaze on the rubber hand and arm, as an experimenter simultaneously and synchronously uses two paintbrushes to touch an analogous location on the rubber hand and his or her real left hand. After a few minutes, nearly everyone who has been subjected to the experiment has reported experiencing the strokes not on the left hand, which received the tactile stimulation, but on the site where they fixated their gaze—the dummy rubber hand. Indeed, most of the subjects reported feeling that during the stimulation, the rubber hand felt as though it was their own true hand.

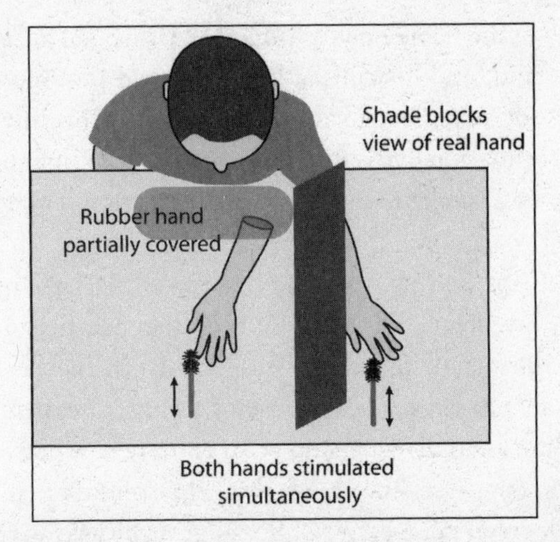

FIGURE 3.4 The brain adopts a rubber hand. The drawing depicts the experimental apparatus used to induce the "rubber hand" illusion. See the text for details. *(Illustrated by Dr. Nathan Fitzsimmons, Duke University.)*

In a subsequent experiment, the same group of subjects endured a longer period of paintbrush strokes. Afterward, they were asked to close their eyes and move their right index finger on the surface of the table until it had reached the index finger of their left hand. Their left index finger reaches were headed toward the rubber hand's finger rather than the index finger of their own biological hands for as long as the rubber hand illusion remained in their head.

If the lasting effects of the rubber hand illusion do not convince you of the dynamic nature of our body image, perhaps an "out-of-body" experience will do the trick. Like phantom limbs, reports of out-of-body phenomena, the vivid perception of exiting the body or even experiencing the body from an outside viewpoint, appear throughout history. Out-of-body experiences can be induced by a number of events, including brain trauma, near-death experiences, car crashes, major surgical procedures, anesthesia with a drug called ketamine, consumption of psychedelic drugs, deep meditation, sleep or sensory deprivation, and sensory overload, just to mention a few. Olaf Blanke of the Brain and

Mind Institute of the Ecole Polytechnique Fédérale de Lausanne, in Lausanne, Switzerland, and his colleagues discovered that multiple aspects of an out-of-body experience can be replicated in healthy subjects by noninvasive stimulation of the junction region of the right temporal and parietal lobes, using a technique known as transcranial magnetic stimulation (TMS).

Building on this finding, Henrik Ehrsson of the Karolinska Institute in Stockholm, Sweden, used a virtual reality apparatus to manipulate visual and tactile signals in healthy individuals. In the experiment, the subjects were able to experience the weird feeling of existing outside of their own bodies, assume ownership of an entirely new body, and "swap" bodies with someone else. To achieve this, Ehrsson first manipulated the so-called first-person perspective of his subjects by asking them to wear a head-mounted display that projected a true stereoscopic image provided by two cameras positioned on the head of a dummy, placed in front of them. The cameras were positioned so that they could offer to the human subjects what would be the mannequin's "first-person perspective," that is, a view of the mannequin's chest and abdomen. Then, in a maneuver reminiscent of the rubber hand illusion, an experimenter stepped between the subject and the mannequin, while staying out of view of the cameras and, hence, out of the subject's view. Using two rods, the experimenter stroked simultaneously the subject's and the mannequin's abdomens for several minutes, keeping the strokes as synchronous as possible. During this stimulation, the subjects were able to see the rod touching the mannequin's abdomen. Strikingly, when asked to describe what they experienced during the experiment, most subjects reported sensing the touch of the rod on the mannequin's abdomen rather than on their own bodies. In fact, the subjects, for the most part, reported feeling that the mannequin's body had become their own. So vivid was this assimilation of the mannequin's body that, when the authors "threatened" to cut the mannequin's abdominal skin with a knife, the human subjects observing the scene through the head-mounted camera exhibited a significant increase in evoked skin conductance response, suggesting that the "threat" to the mannequin's body spawned a great deal of anxiety in them. Similar effects were provoked using other parts of the mannequin's body, such as the hands. However, if objects that did not resemble a human body were utilized, the out-of-body experience did not take place.

Using the same apparatus, Ehrsson and his research team went further, demonstrating that people were able to swap bodies with another person through an out-of-body experience created by another manipulation of visual and tactile information. In the experiment, the subjects wore the same head mount. This time, however, the images the subjects saw were generated by a camera positioned on the head of the experimenter, who sat right in front of the individuals and looked directly at them. This clever arrangement allowed the subjects to view their own hands from the "first-person" perspective of the experimenter. Then, each subject was asked to stretch out his or her right arm and shake hands with the right hand seen in front of them. At this moment, the subject could see and recognize his or her own right arm moving toward the right hand of the experimenter. Because the head mount allowed the subject to assume the experimenter's perspective, when the experimenter moved his or her right hand, the subjects saw this movement as coming from the right side, as well. The subject was then asked to squeeze hands synchronously with the experimenter for a couple of minutes. When asked to describe what had been felt during these two minutes, the subjects predominantly reported feeling that the experimenter's arm, not their own, belonged to their bodies. They also said that their bodies were located behind and to the left of the experimenter's arm, not in front of the experimenter. Their real bodies had been forgotten.

To make things more persuasive, the experiment was repeated while a second experimenter used a knife to threaten either the experimenter's or the subject's arm while the two shook hands. Remarkably, subjects experienced a much higher increase in evoked skin conductance when the experimenter's right arm, and not their own real arm, was threatened by the knife. Interestingly, neither the transfer of body ownership to a mannequin, nor the swapping with another human body, was influenced by gender. Men could swap bodies with women and women could swap bodies with men, a finding that demolished a few neurobiological dogmas. So much for the cult of the body!

The rubber hand illusion and the laboratory induction of out-of-body experiences indicate that the brain actively shapes the sense of self and a bounded physical existence. At the center of this new view of body image is the fact that nothing in our regular daily experience and perception of our bodies, throughout our entire lives, resembles the distorted somatotopic maps traced as a result of Penfield's neurosurgical recordings. If anything, those maps are as strange to us in real life as they look in print. Indeed, when physiological recordings are carried out in behaving animals, the map of the sensory body obtained is vastly more dynamic than the homunculus; even the most precise stimulus to the body's periphery produces a spatiotemporal wave of neuron activation that quickly and widely spreads across the primary somatosensory area, called S1, and other cortical fields. Such experimental evidence, now reproduced by many laboratories worldwide, directly rebuts the mosaic view of the brain built by Brodmann's cytoarchitectonics. And in order to achieve the multimodal integration required to define an internal body image, the brain has to recruit widely distributed neural networks dispersed across most of the neocortex, not to mention the subcortical territories, whose contribution to the sculpturing of a coherent sense of body ownership remains mostly unexplored.

As we will see later in chapter 9, the definition of a body image does not seem to end at the limits of the last layer of epithelial cells that cover our fragile primate bodies. Instead, a series of studies suggest that as monkeys and humans acquire proficiency in the use of artificial tools, their brains assimilate these tools as true and seamless extensions of their own biological bodies. That implies that part of the process of becoming an exceptional violinist, pianist, or soccer player involves the gradual incorporation of specialized tools of the trade, such as violins, pianos, and soccer balls, as add-ons to the neuronal representations of fingers, hands, feet, and arms that exist in the brain.

But it is not just the brains of virtuosos and world-class athletes that are capable of such mesmerizing tricks. In each of us a brain is constantly at work, frenetically assimilating everything that comes into our proximity and morphing our self-image based on a ceaseless flow of information. Our primate brains have the distinguished and unique capability of being not only one of the few, and certainly the most sophisticated, toolmakers bred by natural selection, but also an equally voracious tool incorporator. Our brains are constantly busy, adding our clothes, watches, shoes, cars, computer mice, silverware, and every other instrument we use into a dynamically expanding and contracting body representation.

Taking these ideas to the limit, such findings and theories support the contention that as we learn to utilize brain-machine interfaces in which our brains interact directly with artificial tools, located either next to or far away from our biological bodies, our brains will incorporate these devices as part of us. For some, the potential future amalgamation between brains and machines may sound frightening and even seal the end of humanity as we know it. I could not disagree more. In fact, I believe that this tool-incorporating hunger of the brain will open a new chapter in evolution, offering us ways to extend our bodies, perhaps reaching immortality, in a very particular form: by preserving our thoughts for posterity.

This neurobiological principle of assimilation may reach even deeper into our lives. Evidence suggests that as we fall in love, it is our brains, not our hearts, that seamlessly merge scents, touches, sounds, and tastes to transform the body of our beloved into a passionate and vivid extension of ourselves. That is why I believe that Cole Porter actually got the big picture right when he wrote "I've Got You Under My Skin."

To hear the "chairman of the board" Frank Sinatra belt out Porter's lyrics during an innocent summer night spent dancing with your Gala in an otherwise empty parking lot is enough to understand how painful it is to be deprived of the touch of a true love. If you had any doubt before, you can rest assured: the pain of love is real, very likely because, for our brains, to lose the object of our love is to sever a real part of our lonely selves.

It is no wonder, then, that millions of people cannot entertain the thought of being separated, not even for a minute, from their beloved BlackBerrys. Once the primordial feelings aroused by the simulated body are released, the brain embraces them without boundaries.

4

LISTENING TO THE CEREBRAL SYMPHONY

I have always been amazed at how most people, including a large number of neuroscientists, readily embrace the notion that the central nervous system needs to be organized in some sort of hierarchical and orderly manner. Somehow, though, this type of rigid, military-like structure appeals to us humans as being natural, "the way things ought to be." Yet, in my opinion, the assumption that the brain follows an operational hierarchy has more to do with ideology than with the way nature really operates. Perhaps it's not so surprising that science historian Philip Pauly traces the localizationist dogma back to a particularly orderly source in his paper, "The Political Structure of the Brain: Cerebral Localization in Bismarckian Germany." Pauly notes how both Eduard Hitzig and Gustav Fritsch, the discoverers of the motor cortex, eagerly borrowed metaphors from their beloved Prussian bureaucracy to elucidate the workings of the central nervous system.

There is no doubt that a hierarchical brain, exulting order from the single neuron up to Korbinian Brodmann's fifty-two cortical areas, was more palatable to most of the neuroscientists who established the field in the late nineteenth and early twentieth centuries. Ideology aside, the limitations of using language may also be responsible for the human tendency to categorize and organize how thinking happens. Certainly naming something in the brain was the fastest road to immortality a nineteenth-century neuroanatomist could take. Robert Erickson, my colleague at Duke, insisted that because we always search for words to define natural phenomena, including the types of behaviors and functions carried out by the brain, we almost imperceptibly make the logical jump that each of these corresponding and discrete "word-function" categories, as Erickson called them, should be represented by a particular brain region. As my research over the past two decades has shown, however, brain activity involves overlapping yet widely distributed neural circuits, interacting across time. Words alone may not do justice to the brain's work, which is much more probabilistic than any of the world's many languages that we use to express our thoughts.

One of the first scientists to recognize the distributed nature of thinking was Sir Charles Sherrington, who proved that basic neurological functions, such as the spinal reflexes, depended on the cooperation of multiple peripheral and central structures of the nervous system, a finding that earned him the other half of the 1932 Nobel Prize in Physiology or Medicine. By defining this neural collaboration as the work of an integrated system, Sherrington helped launch the area of brain research now called systems neuroscience. Sherrington was not a man purely devoted to the laboratory bench. In his book *Man on His Nature*, he captured the brain's inner life in magnificent, poetic prose: "The brain is waking and with it the mind is returning. It is as if the Milky Way entered upon some cosmic dance. Swiftly the head-mass becomes an enchanted loom where millions of flashing shuttles weave a dissolving pattern, always a meaningful pattern though never an abiding one; a shifting harmony of subpatterns."

Sherrington's integrative neuroscience rekindled the use of a distributed strategy to describe the physiology of thinking. But as often happens in science, the technology available for investigating the global function of the brain was meager, to say the least, in the opening decades of the twentieth century when Sherrington's work made its mark.

This started to change in 1924, when Hans Berger, a German physician and professor at the University of Jena, made a stunning discovery. Berger had at one time been a collaborator of Oskar Vogt and Brodmann, the paramount cytoarchitects. Upon his return from World War I, Berger had become frustrated by his inability to use measurements of the blood circulation in the brain to establish a link between neural activity and psychological behavior. So he decided to venture in a different direction: to measure the brain's electrical signals. Richard Caton, an English scientist, had recorded the electrical activity produced by the brain from the exposed cortices of experimental animals back in 1875, and the Russian physiologist W. Prawdicz-Neminski had recorded similar activity from the intact skulls of dogs just before the war. Berger decided to utilize the same technique on humans. After first trying to make recordings by inserting silver-wire electrodes into a person's scalp, Berger realized—to the relief of his subjects (including his own son)—that he could obtain a reading on the small electrical potential generated by the human brain simply by attaching silver-foil electrodes on the scalp and connecting them to a galvanometer. Berger proceeded to record brain activity in a variety of conditions, including in patients suffering from epilepsy, and learned that a series of global brain rhythms were associated with routine behaviors. One of his first discoveries was the alpha rhythm, an oscillation of ten cycles per second (or ten hertz) in brain potentials recorded over the occipital bone when patients sit quietly and immobile with their eyes closed. Berger named his method the electroencephalogram, or EEG (see Fig. 4.1).

Today, the EEG is routinely employed as an essential diagnostic and research instrument. Moreover, with the discovery of the EEG, neurophysiologists began to realize that the cortex was capable of producing global patterns of electrical activity, including a wide range of rhythms

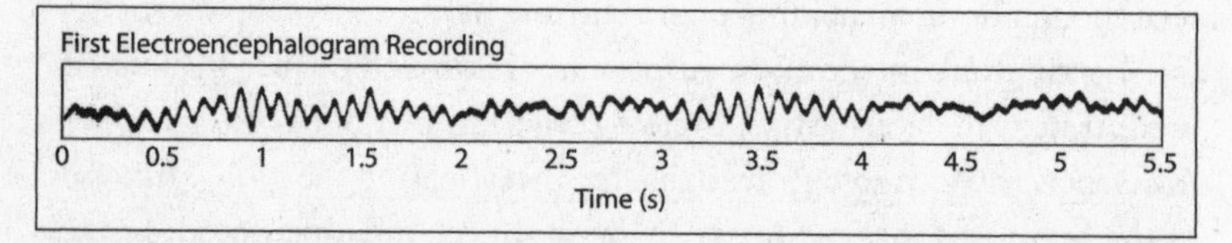

FIGURE 4.1 Sample of first EEG recording ever obtained by Hans Berger. The trace represents a few seconds of the electrical activity of Berger's own son, recorded using scalp sensors. *(From Hans Berger, "Über das Elektrenkephalogramm des Menschen,"* European Archives of Psychiatry and Clinical Neuroscience *87, no. 1 [1929]: 527–70, with permission from Springer.)*

that were correlated with a variety of normal internal dynamic brain states and behaviors, such as attentiveness or relaxed wakefulness. Pathological cortical conditions, including different types of epileptic seizures, could also be detected in EEG records. The EEG gave neurophysiologists the ability to document the integrative activity of an awake brain.

Despite the development of EEGs and another method, the sensory evoked potential, to measure cortical electrical activity triggered by a sensory stimulus, the tradition of "looking locally" for brain functions continued unrepentantly as neuroscientists, led by Lord Adrian and his Cambridge colleagues, simultaneously mastered techniques for recording the electrical signals, also known as action potentials, produced by individual neurons. Their tool of choice was the microelectrode (see Fig. 4.2). In its most traditional incarnation, a microelectrode consists of a long, thin, rigid metal rod with a very sharp tip. With the exception of its tip, the entire microelectrode shaft is encased in an isolating material, for example, a resin, glass, or plastic. Once the surface of the brain is

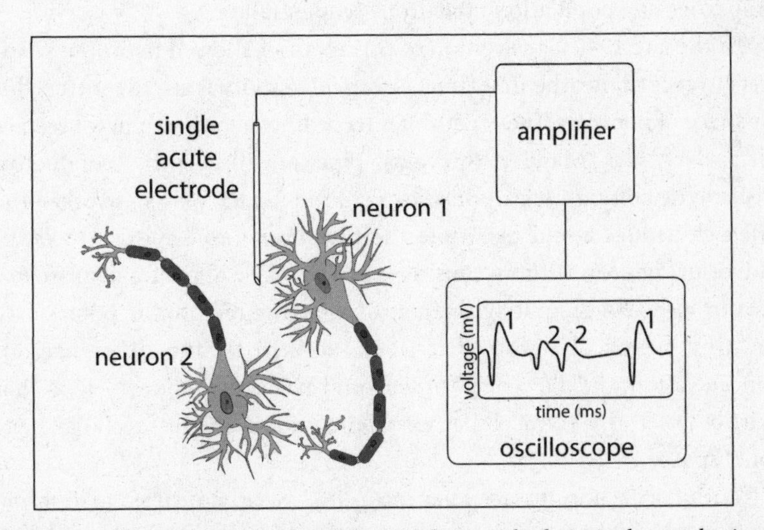

FIGURE 4.2 The single microelectrode recording method. A single metal microelectrode, positioned in the extracellular space next to two neurons, is capable of recording the extracellular action potentials of both cells. Inspection of the recording trace in the oscilloscope reveals that the action potentials of the two neurons have different shapes and magnitudes, allowing the distinction between the two cells. *(Illustrated by Dr. Nathan Fitzsimmons, Duke University.)*

exposed, a single microelectrode can be lowered into the brain tissue, creating a penetration track along which one can record the electrical activity of a single neuron in the extracellular space where the microelectrode's tip rests. A ground wire, usually connected to the dura, is used to provide a reference point for the electrical signals that are picked up. Since the extracellular electricity generated by neuronal action potentials is very small—around one millivolt—the signals registered by the microelectrode must be filtered and amplified in order to record the activity of a given single neuron.

Using a microelectrode, neurophysiologists can monitor the activity of a neuron for a period spanning from a few minutes to, at most, a couple of hours. Once a neuron's action potentials are isolated and recorded, the microelectrode can be moved a few micrometers deeper into the brain to record the activity of another cell. Neuroscientists use such a serial sampling procedure, which can be repeated several times during a recording session, by advancing or retracting the microelectrode or by initiating new tissue penetration in another area, to characterize the properties of a population of neurons sequentially.

In the late 1940s, a variation of this method allowed neurophysiologists to record, for the first time, electrical activity from the intracellular space of neurons. These neuronal recordings required a new breed of microelectrodes. Made of thin glass pipettes, filled with a conductive electrolytic solution (e.g., potassium chloride), and very sharp tips, the microelectrodes could penetrate the membrane and cytoplasm of an individual neuron without causing much damage, allowing neuroscientists to measure accurately the neuron's resting membrane potential—the tiny electrical dipole that is maintained by the differences in concentration and flow of positively and negatively charged ions that exist between the extracellular environment and the intracellular neuronal space.

Within a decade, intracellular recordings were being used to describe the myriad of synaptic currents that continuously bombard the neuron's membrane and that can lead to the production of action potentials. Once triggered, action potentials propagate very rapidly to the neuron's axon and through all its branches, which make synaptic contact with other neurons in a brain circuit. That is how the firing of a given neuron can influence the physiology of every neuron connected to it.

Much of the pioneering experimental work on synaptic potentials and their role in generating action potentials was conducted by one of Sherrington's former students, the Australian neurophysiologist and Nobel laureate John Eccles. Taking full advantage of a new method for in vivo cellular recordings, Eccles discovered that synaptic potentials, generated by the stimulation of distinct peripheral nerve fibers that projected to neurons in the spinal cord, could exert either excitatory or inhibitory influences on the membrane potential of those neurons. He also observed that if the overall sum of these synaptic influences crossed a particular voltage threshold, a spinal cord neuron would respond by producing an action potential.

Armed with their beloved microelectrodes, for two decades neuroscientists conquered a variety of brain structures employing Adrian's experimental approach, with a fervid devotion to the primary sensory fields that had been so instrumental in validating Brodmann's cytoarchitectonics. One of the main protagonists of that era, the legendary American neuroscientist Vernon Mountcastle, made his name wielding microelectrodes across the primary somatosensory cortex (S1) of anesthetized cats and monkeys. During a series of painstaking experiments, reported in the *Journal of Neurophysiology* in 1957, Mountcastle and his students at Johns Hopkins University penetrated the surface of the S1 cortex with a microelectrode and serially recorded the tactile responses of the individual neurons they encountered as they advanced the microelectrode down into the depths of the brain tissue. By showing that most of the neurons they encountered during each of these penetrations shared similar physiological properties—for instance, the neurons would all fire when the same skin region was stimulated mechanically—Mountcastle provided physiological evidence for the hypothesis that S1 was formed by a series of functional columns, a "vertically linked group of cells," which he proposed constituted the "elementary unit for cortical function."

In his classic paper, Mountcastle took proper care to acknowledge that this type of vertical cortical modular organization had previously been proposed by the prodigious Spanish neuroanatomist Rafael Lorente de

Nó. At the age of twenty, de Nó had kicked off his career by picking a fight—a scientific one—with none other than Don Santiago Ramón y Cajal, by submitting to one of Cajal's magazines an article that argued forcibly, contrary to Cajal's opinion, that the organization of the mouse cortex was virtually identical to, and as rich as, that of the human cortex. Rather than balking, Cajal published the article, although the relationship between both Spaniards was strained. Sadly, Mountcastle's painstaking footnote was, for the most part, overlooked by the many scientists searching to uncover the mysteries of the mammalian cortex.

Mountcastle was far from alone in his physiological mission. The industrious Canadian-Swedish partnership formed by David Hubel and Torsten Wiesel was also busy recording single neurons in anesthetized cats, but in the primary visual cortex (V1) rather than S1. Using the same serial sampling technique, Hubel and Wiesel markedly expanded on the work of their mentor, the American neuroscientist Stephen Kuffler. Kuffler had mapped the receptive fields (RFs) of retinal ganglion neurons, which are positioned near the surface of the retina of the eye, and had found them to be nearly circular.

Yet, when Hubel and Wiesel sampled single V1 neurons, they discovered that each of them primarily responded to a bar of light that was presented at a particular angle or orientation. Other V1 neurons responded more strongly if the bars of light moved in a particular direction. Hubel and Wiesel proceeded to identify the orientation preference of a sequence of neurons that they penetrated in the V1 of cats, creating a complete map of these preferences. In the map they spotted a large number of cortical columns, each of which contained, throughout its vertical depth, neurons that gave a similar maximal and specific firing response when a bar of light of a given orientation hit the eye. Hubel and Wiesel's V1 map provided substantial ballast for the theoretical framework defended by Horace Barlow, a neuroscientist at Trinity College, Cambridge. The great-grandson of Charles Darwin, Barlow promoted the notion that single neurons worked as "feature detectors," responding to specific components of a complex stimulus.

■ ■ ■

From the very earliest recordings of single neurons made by Lord Adrian in the 1920s to the breakthrough experiments conducted by Hubel and Wiesel to the heyday of the technique in the 1980s, the basic experimental approach used to investigate how the brain generated perceptual experiences did not evolve much. For starters, perceptions were always investigated from the perspective of an external observer who delivered a well-controlled, unimodal stimulus and targeted it at a particular peripheral receptor (e.g., the skin, retina, inner ear, or tongue) of a subject animal. The experimenter was also responsible for measuring the evoked reaction produced by the animal's brain and controlling the brain's internal state. That explains why, in the vast majority of the studies, animals were kept in a deep state of anesthesia. Experimental conditions were easy to control that way.

This experimental approach biased neuroscientists toward the notion that the brain rests static, devoid of any historic recollections and patiently waits to decode the information embedded in external physical stimuli by breaking down these messages into their constituent features. In the case of a visual stimulus, for example, these compartmentalized features include orientation, color, and motion. That reductionist paradigm removed any possibility of incorporating the brain's internal point of view, which in even a simple, single-stimulus experiment involves:

- its internal dynamic state at the moment of its encounter with the stimulus, as well as the internal expectations created by the brain just prior to the encounter;
- the accumulated evolutionary and individual perceptual history of the subject, which summarizes the brain's multiple previous encounters with similar and dissimilar stimuli;
- its adaptive ability, which allows it to change in response to an encounter with a novel perceptual experience;
- the emotional value associated with the stimulus; and
- the brain's production of a series of motor behaviors, including eye, hand, and head movements, aimed at actively sampling the stimulus.

Instead, experimenters focused on identifying neurons that could be driven to fire maximally in response to a primitive stimulus, and the notion took hold that the individual features of the stimulus would trigger only a particular and specialized group of cortical neurons, located in discrete cortical areas. Gall would certainly have been proud of this next generation of localizationists. In short order, cortical neurons were being categorized by their theoretical "feature-extracting" capability. These included, of course, Hubel and Wiesel's specialized V1 neurons for detecting oriented lines. And many other neurons were joining them, with cascading effects.

Cells for detecting color or motion were discovered in cortical visual fields other than V1, leading to the establishment of a strict visual cortical hierarchy that originated in V1 and formed distinct streams of visual processing, known as the dorsal and the ventral pathways through the occipital, parietal, and temporal lobes. Because it was presumed that the location of the neurons in these two streams determined their roles in visual processing, the two systems were initially dubbed the where/how (dorsal) and what (ventral) pathways of the visual system. Further investigation, however, demonstrated that there was significant cross talk between these streams. Today many question the usefulness of this scheme.

Similar segregated functional pathways were proposed for the auditory system and, with less emphasis, the somatosensory and motor systems. Yet, none of these attempts could rival the grandiose schema built by the physiologists probing the parietal, temporal, and occipital lobes where the visual cortex was said to reside.

Through the 1980s and into the early '90s, there were few souls brave enough to confront Gall's inheritors. Systems neuroscientists almost unconsciously fell into the trap of classifying each and every bewitching neuron, drawing their names from a set of proverbial, and sometimes anthropomorphic, labels. For instance, a neuron that vigorously responded to the flash of a familiar face—say, the picture of the person's grandmother—was baptized, in 1969, as the "grandmother neuron." The homely grandmother neuron was joined by a neuronal celebrity in 2005. That was when the electrical discharges of the "Halle Berry neuron," which responded "to the concept, the abstract entity, of Halle Berry," were found in the inferotemporal cortex of a male patient and recorded in all their splendor at the University of California, Los Angeles. Where else would a celebrity neuron make its first appearance?

At first glance it probably seems natural that the activity of an individual, feature-tuned neuron would play a starring role when the image of such a potentially beloved and admired lady as your grandmother or Academy Award winner Halle Berry enters your visual field. In reality, however, these names merely identify the best visual stimulus that experimenters were able to utilize to get a single neuron to fire maximally, and some of the recorded neurons also responded, albeit at lower intensity, to other visual stimuli. Yet, by focusing only on one "high-profile" stimulus, neurophysiologists became more and more fixated on the target of their microelectrodes' measurements: the single neuron. As one of the classiest American neurophysiologists and historians, James T. McIlwain of Brown University, put it: "The widespread use of microelectrodes focused experimental research on the behavior of single neurons and the possibility that their individual properties could account for much of what the brain does. . . . I can testify personally to the seductiveness of this . . . view. As you sit in a darkened laboratory with your attention riveted to the sounds of the audiomonitor and probe a neuron's receptive field with a tiny visual stimulus, it is easy to forget that the cell you are listening to is but one of many that are responding to the stimulus."

Notwithstanding the sheer, simple elegance—and seeming experimental success—of labeling the physiological properties of single neurons, in the 1980s the German computer scientist Christoph von der Malsburg exposed a fundamental limitation of the feature-extraction model. Malsburg's challenge, which became widely known as the binding problem, goes like this: If the brain really treats a novel sensory stimulus by first decomposing its overall complex structure into a series of discrete, primitive features, each of which is represented by a single specialized group of neurons located in a given cortical area, how can the brain put back together all this information, which has been broken down (by feature) and distributed (spatially) across the cortex to reconstruct the original stimulus and generate the full perceptual experience of a complex object that we know from daily life?

That was a good question!

The proponents of the feature detector theory of single neurons had no immediate answer for Malsburg. It was as though he had opened a chasm in the field, much like the one between general relativity and quantum mechanics. Yet, after the initial furor provoked by Malsburg's inconvenient but rather pointed challenge, most neuroscientists returned to using the same old techniques, the same old terms, and the same old thinking.

Since the early 1950s, there have been a few mavericks willing to stray from the plodding of a single microelectrode into more adventurous attempts to sample the activity of populations of neurons. None of these early dissenters was more daring than the American neuroscientist, philosopher, and writer John Cunningham Lilly.

After graduating, on an academic scholarship, from the California Institute of Technology in 1938, Lilly earned a medical degree from the University of Pennsylvania, where he also trained in psychoanalysis. After World War II, he landed at the National Institutes of Health in Bethesda, Maryland, as the chief scientist of the exotically bureaucratic section called Cortical Integration. Lilly's years at NIH gave an early indication of a life committed to exploring unconventional ideas. Over the subsequent five decades, he was associated with a number of unusual and often highly controversial research initiatives, several of which raised eyebrows.

Lilly had a long-standing interest in human consciousness. In 1954, he plunged into an investigation of how the human brain would react to an environment devoid of sensory stimulation. In his quest he designed and extensively used an apparatus he named the sensory isolation tank. Lilly and one of his friends, Edward Evarts, served as both the first subjects and the principal investigators. In the original studies, which later inspired the Hollywood film *Altered States*, starring William Hurt as a scientist devolving into a blob of consciousness, Lilly and Evarts took turns lying inside the tank, which was soundproofed and contained slowly flowing, warm salt water. Their bodies were suspended so that only the top of their heads remained above water. A mask covered their heads to further reduce sensory stimulation. While underwater, normal breathing was facilitated by attaching a tube to the mask. After a few training exercises to get accustomed to the conditions, Lilly and Evarts experimented with the tank. They would be isolated inside for a couple of hours, during which they were supposed to relax and restrict their body movements. At the end of each session, they would file a report of their impressions of the experience. Following a decade of self-experimentation with the isolation tank, Lilly expanded his daredevil research by placing himself in this environment, either alone or in the company of live swimming dolphins, animals that he was very fond of, after taking the psychedelic drug LSD.

Lilly's experiments with LSD and other drugs, as well as his attempts to demonstrate possible ways to establish direct communication between humans and dolphins (one of his most exotic ideas), strained his relationship with more orthodox scientists. As a result, his more mundane, albeit revolutionary neurophysiological studies have all but disappeared from the neuroscience literature.

Starting in 1949, Lilly and a few coworkers set out to find a new method for recording large-scale electrical brain activity and for chronically stimulating brain regions with electrical pulses without producing tissue damage in the process. Lilly's goal was to unite neurophysiology and experimental psychology, which remained completely balkanized in academia. In his view, until the two fields came together, it would be impossible to generate "[an] accurate time-space description of central nervous system electrical activity and behavior in the very short-term intervals." The first step would be to probe the brains of animals that had not been anesthetized.

Ever the pioneer, Lilly built an experimental setup capable of implanting electrodes in unanesthetized animals, adding to this new paradigm the capability of recording the electrical activity of the brain. And instead of limiting himself to only a single electrode, which could be placed in a given brain structure to record the electrical signals of a sequence of neurons over time, Lilly designed a multi-electrode array that allowed him to sample simultaneously electrical potentials of more than two dozen spots on the cortical surface. (His subjects in this experiment were cats and monkeys, rather than himself.) Lilly named his apparatus the "twenty-five-channel bavatron" and the brain electrical profiles he obtained with it, "electro-iconograms."

To render data from behaving animals, in his (mostly) precomputer age, Lilly had to overcome enormous technological bottlenecks. In an early version of the bavatron, he threaded twenty-five metal wire electrodes into an array of twenty-five glass tubes, arranged in a five-by-five square, with two millimeters of space between each electrode. Each of the glass tubes was embedded in a Lucite cylinder, and the cylinder was mounted in a stainless steel barrel; the barrel was screwed into a hole, three-quarters of an inch in diameter, drilled into an animal's skull. During an experiment, the animal was housed in a soundproof box that itself was placed in a room shielded with metal, in order to reduce the amount of stimulus disruption received by the animal and the amount of electromagnetic "noise" generated by sources outside the animal's brain—including radio broadcasts from Washington, D.C. To further minimize noise, the glass tubes were filled with a saltwater solution and made to rest gently on the cortical surface. Once the barrel and the electrodes were in place, each of the wires was connected to a twenty-five-channel preamplifier situated next to the animal. The outputs obtained by this preamplifier were fed via long cables to twenty-five amplifiers located outside the shielded room. A scientist who worked at NIH for many years once told me that these constituted the entire stock of amplifiers available at the time to the whole staff of the National Institutes of Health.

Lilly certainly was up to something big.

From each of the twenty-five channels amplified and filtered by his array, Lilly recorded the difference between the potential obtained from a given electrode and the average potential obtained from the full array. This design permitted signals that were common to all of the electrodes in the array to be canceled out from each individual electrode recording, allowing only the relevant local electrical brain activity to be isolated from each sensor. This technical innovation, known as differential recording, is still used today to remove movement artifacts and other strong biological signals from neurophysiological recordings in behaving animals.

Lilly's insight and creativity reached their pinnacle in his system for capturing and recording the spatiotemporal waves of brain activity with the twenty-five-channel bavatron. Without the luxury of computers or other large data-storage devices, Lilly linked each of the outputs of his twenty-five amplifiers to a twenty-five-square array of glow tubes, disposed into the same five-by-five arrangement as the twenty-five electrodes resting on the cortical surface. In this amazing apparatus, the intensity of the light emitted by each glow tube was modulated, either above or below an averaged value, by the differential electrical signal generated by the corresponding electrode in the array connected to a given bulb. Thus, if the electrical signal coming from a given electrode was negative (in relation to the mean potential of the full array), the glow lamp would brighten. Conversely, if the signal of one electrode was positive, the light emitted by its corresponding glow tube would dim. Using this strategy, Lilly began to observe spatiotemporal waves of light that corresponded to spatiotemporal patterns of electrical brain activity being recorded from a particular spot on the cortical surface while his animal subject performed some type of movement, listened to auditory stimuli, or simply went to sleep and then awoke. Through the sort of boundless, timeless link that only human endeavors, like science and art, can establish, John Lilly, secretly sequestered in a lab in Maryland, became the first neurophysiologist to glimpse the dynamic blossoming, spreading, and vanishing of unabiding, dissolving neuronal patterns as they were carefully woven by an enchanted mental loom, much like the one envisioned, five decades earlier, by the great Sir Charles Sherrington.

As if this was not enough wizardry, in order to generate permanent records of these complex spatiotemporal patterns of cortical activity, Lilly employed an electrically driven, sixteen-millimeter Bell and Howell 70 G super-speed motion picture camera to tape continuously the patterns of light produced by his glow-tube array. In one of his papers describing this method, Lilly implies that the camera made so much clatter that he had to find a spot for it in the building as far as possible from the shielded room with his animals so as not to distract them with the noise of making a "brain film." (Later, Lilly would devise a method for converting individual frames from his brain films into careful woodcarvings: solid renderings of fluid brain dynamics.)

In the early days, Lilly carried out his experiments on anesthetized cats. That provided a simple test of the electrode array and the entire recording apparatus. He shared these findings in a series of papers published in leading neurophysiological journals. During this period Lilly also introduced a new method for stimulating the brain chronically, using a biphasic, charge-balanced electrical pulse that did not damage tissue, a technique sometimes still referred to as a Lilly wave. Eventually, Lilly progressed to experiments involving awake, behaving monkeys, whose heads were restrained by the recording apparatus though they could either freely move their upper and lower limbs or attend to sensory stimuli, such as a click sound. Unfortunately, much less is known about the results he obtained in these studies. In a single book chapter, Lilly briefly describes one of his most elaborate experiments—implanting an astounding 610 cortical electrodes across most of the cerebral cortex of an adult rhesus monkey. Since Lilly's apparatus limited him to recording only 25 electrodes, he was never able to collect the simultaneous brain activity from all 610 sensors.

Although in the last years of his life he may not have been aware of (or cared about) the evolution of the field he helped to create, it is fair to say that Lilly's neurophysiological experiments marked his life deeply. Indeed, while writing about his time in the isolation tank, Lilly recalled that during his first LSD trip he experienced the feeling of navigating within his own brain, of seeing his neurons firing. Those long nights at NIH making brain films left enduring memories in Lilly's brain.

During the 1960s, Edward Evarts, who collaborated with Lilly in the isolation tank studies, introduced his own method to perform brain recordings in awake and behaving nonhuman primates. His approach, which became the gold standard in primate neurophysiology, focused on single neurons that were recorded while a monkey performed a particular behavior task. Most of what is known about the physiological properties of individual cortical and subcortical neurons in primates has been gathered using this method or variations on it. Its success is undisputable. Yet, without a multi-electrode array, neurophysiologists cannot understand how *populations* of neurons in a brain circuit operate in real life—and that poses several significant limitations.

For starters, as we have seen, until very recently these single neuron recordings were obtained using a single microelectrode that was slowly moved through the depths of a particular cortical or subcortical structure so that electrical activity of individual neurons could be recorded, one at the time, in a sequential way. Second, in the traditional version of the experimental setup, these single-unit recordings start only at the end of the animal's behavioral training. By the time any neuronal activity is recorded, the animal is overtrained in the task of interest. This leads to the confounding problem that by encountering single neurons whose firing properties correlate to particular task contingencies imposed by the animal's behavior training, an experimenter cannot decipher whether these findings reflect intrinsic physiological attributes assigned to these neurons or simply depict the fact that these highly adaptable neurons have been conditioned to fire according to the salient features of the task. In some cases, neurophysiological experiments seem to have morphed into tautological exercises, in which authors report, usually with great excitement, that after many months of intense behavioral training and painstaking analysis of neuron firing patterns, they were capable of identifying a few neurons whose activity correlates with some of the task's main attributes. Given that the behavioral task was usually the most relevant activity carried out each and every day by the animals studied, this is not at all surprising. The much tougher question to answer is whether the neuronal firing is *causally* linked to the execution of the task. In Evarts's paradigm the focus is also often placed on sampling neuronal activity in a single brain structure. Because of the technological constraints imposed on experimental research, Sherrington's integrative brain was once again sliced into its multitudinous parts and microscopic constituents, rather than being allowed to express its richness as a whole complex neural circuit.

Yet, few believed there was a way to break this traditional approach to probing the brain.

One afternoon circa 1987, during a break from an experiment that was not going well, I found myself browsing through a book that I had recently purchased at my favorite bookshop in downtown São Paulo. The book, written by the Savilian Chair of Astronomy at Oxford University, Joseph Silk, was titled *The Big Bang: The Creation and Evolution of the Universe*. As I distractedly flipped page after page, I fell upon an illustration that grabbed my attention: a three-dimensional depiction of radio wave sources obtained by a milelong radio telescope formed by placing smaller radio telescopes in phase at a site in England (see Fig. 4.3). By plotting the magnitudes of the radio sources as *Z*-axis peaks distributed

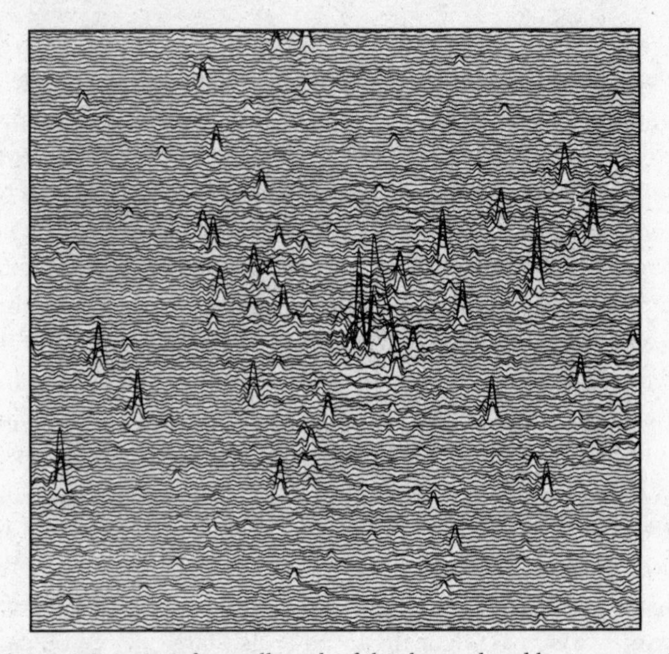

FIGURE 4.3 3-D image of a small patch of the sky produced by an array of radio telescopes in Cambridge, England. Peaks identify galaxies that emit radio signals. The height of each peak indicates the magnitude of the radio signal produced by each galaxy. *(Originally published in Joseph Silk,* The Big Bang *[San Francisco, Calif.: W. H. Freeman, 1980], with permission from Mullard Radio Astronomy Observatory [MRAO] and Cavendish Laboratory, Cambridge.)*

on a two-dimensional space represented on the *X* and *Y* axes of the plot, the resulting "map" identified the location of radio galaxies and quasars in a sector of the universe. The more I inspected the plot, the more I felt that the same approach could be applied to studying brain activity.

Spurred by Silk's book, my idea was to implant several "sensors" in multiple locations of a given brain circuit, to create a similar 3-D neurophysiological map. Furthermore, if a fourth dimension, representing time, could be added to the map, I would have devised a completely new way to visualize, monitor, and measure the electrical activity of whole brain circuits in behaving animals.

Weeks later, I mustered enough courage to talk to my mentor, Dr. César Timo-Iaria, about my neurophysiological chart work. It did not take long for him to inspect my humble sketches and peruse the couple of pages that timidly and naively outlined a new experimental approach in systems neurophysiology: to watch the brain's work from the brain's own perspective.

"I think it is time for you to finish your thesis, leave the laboratory, and go abroad." His verdict was swift and blunt.

"Have I done anything wrong?" I could barely believe his reaction.

"Nothing at all. You are simply ready to go. And what you want to do, neither I, nor anyone else in Brazil, can help you achieve."

Fifty application letters and ten months later, I found myself one afternoon contemplating a large welcome envelope from a young associate professor, John Chapin, at Hahnemann University in Philadelphia. Inside, I found a series of goodies. The first was a detailed research plan, funded by a grant from the NIH, for creating a new neurophysiological method that might one day produce the spatiotemporal map of the brain I had in mind. Original and daring, the plan involved a level of technical innovation I had not encountered in any of the dozens of scientific papers I had read in graduate school. Chapin aimed to move neuroscience away from single-neuron recordings to a new technique that would allow for simultaneously monitoring populations of individual neurons in behaving animals, not just for a few hours but for several weeks, or even months, at a time. He was reaching for the holy grail of neurophysiology, and in the next five years!

At first glance, every step of the project looked like an impossible, borderline-crazy endeavor. For instance, instead of employing rigid metal microelectrodes, the classic tool utilized by neurophysiologists for almost half a century, Chapin proposed to take advantage of a new type of recording sensor, built in the format of arrays or bundles containing eight or sixteen hairlike, flexible microwires composed of stainless steel. For electrical isolation, each of these microwires would be encased in a thin layer of Teflon that covered all but the microwire's blunt tip. When implanted in the brain, Chapin was certain that these devices would allow us to monitor tens of neurons simultaneously in anesthetized or freely behaving rats alike. To achieve this feat, the microwire arrays or bundles would need to be implanted in a given brain structure, such as the rat somatosensory cortex, during what looked like, even after several of my readings, a pretty demanding neurosurgical procedure. Contrary to the norm at the time, the entire recording apparatus would remain implanted—permanently—in the animal's brain. That was quite a departure from the status quo.

Previously, some neuroscientists had tried to leave rigid, sharp-tipped microelectrodes in the brain as an attempt to record the activity of neurons for longer periods of time. Virtually all of these attempts had failed miserably. After a couple of days in the brain, the microelectrodes mostly stopped working and no neuronal signals could be recorded from them; the inflammatory reaction triggered by implanting a foreign object in the animal's brain produced deposits of proteins and cells along the entire microelectrode surface, particularly at its sharp and bare tip. These deposits blocked neuronal electrical signals from reaching the only exposed area of the sensor. Moreover, as the brain moved slightly inside the animal's head, the microelectrode's rigidity tended to create, over time, large tissue lesions that led to the degeneration of the neurons lying next to the sensor.

Chapin believed both these issues could be solved, or at least mitigated, by using flexible microwires with blunt tips (see Fig. 4.4). Because of the larger exposed area of the microwire tips, he postulated that the inflammatory reaction would not be sufficient to block a tip's entire surface. Instead, the inflammatory deposits would actually improve the quality of the microwire's recordings over the first two weeks or so, morphing it from an electrode with low to one with high resistance. The reaction would also anchor the filament to the brain. And because the microwires were rather flexible, once they became anchored they would move in synchrony with the brain and therefore induce no significant brain lesions.

During the surgical implantation, Chapin's microwire arrays would be pushed gently into the brain tissue so that their exposed blunt tips would rest in the extracellular space surrounding many neurons. This arrangement would allow action potentials produced by many individual neurons to be recorded continuously by multiple amplifiers. In his proposal, Chapin hinted that a completely new piece of hardware would have to be built to handle the overwhelming amount of data recorded in

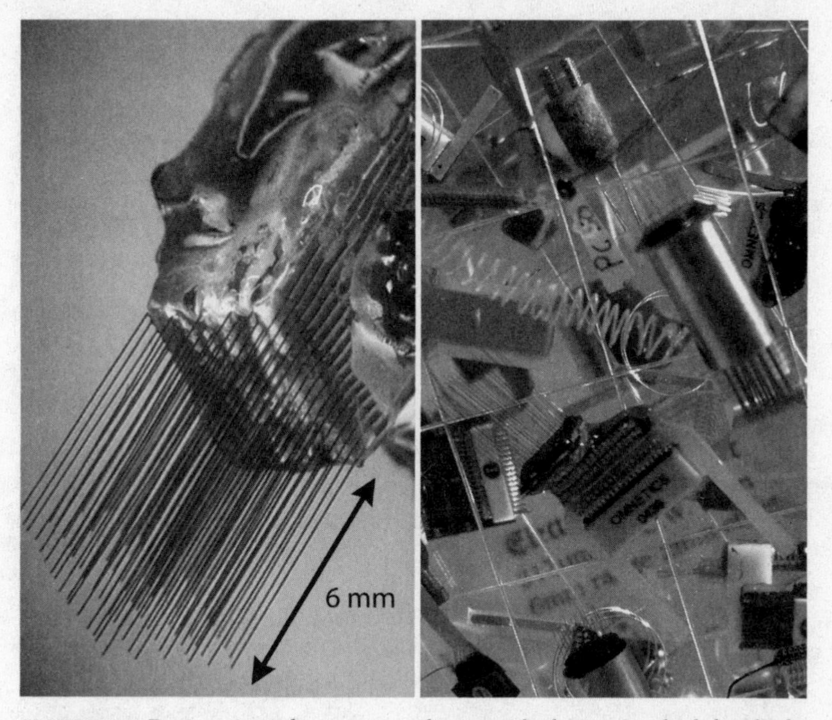

FIGURE 4.4 Engineering a better way to listen to the brain. On the left, a high-power magnification of a multi-electrode array produced at the Duke University Center for Neuroengineering (DUCN) by Gary Lehew and Jim Meloy. Notice that multiple thin metal filaments are clustered in a matrix format. Such filaments are flexible and can be chronically implanted in the brain, remaining active for months to years. On the right, a sample of different types of multi-electrode arrays created at the DUCN in the past decade. *(Courtesy of Dr. Miguel Nicolelis.)*

the experiments. He also realized that a variety of new analytical techniques would have to be developed to make sense of the recorded data.

Based on his preliminary findings, which were published as part of his doctoral dissertation, Chapin noted that rats recovered quickly from this neurosurgical procedure. In fact, they went on to live normal rodent lives after the microwire arrays were implanted. When they returned to their normal routines, doing everything a regular rat might do in the lab, including mastering elaborate behavioral tasks, he could record their brain activity.

Reading the NIH grant, I quickly grasped the enormous influence that the experiments Chapin was proposing might have on the field of neurophysiology if they were to truly succeed. For the first time, there was a scientific road map to the most unknown frontiers of the brain: the place where the electrical storms of distributed neuronal populations combined into streams of thoughts. Getting there would be very difficult, perhaps impossible. But it was a trip worth attempting, one filled with pure adventure.

I arrived at Hahnemann in 1989 as a postdoctoral fellow. There, as the newest apprentice of the Merlin of multi-electrode recording, I joined Chapin in his obsession with a much more mundane challenge.

Both of us desperately wanted to know how in heaven rats manage to escape from cats.

5

HOW RATS ESCAPE FROM CATS

As the computer-controlled sliding doors instantaneously opened, revealing the pitch-dark yet familiar chamber, Eshe did exactly what was expected of her after all those demanding weeks of training. Without hesitation—and likely counting on the reward she was certain to receive given her superb performance of late—she lunged inside the narrow room at full speed, aiming for the opposite wall.

Clearly, she was ready to show off.

The experimental trial started at once, as Eshe's head crossed an infrared light beam located in front of an aperture positioned directly in her running path. The aperture, formed by the small arms of two T-shaped metal bars, each of which protruded from the lateral walls of the chamber, defined an opening through which Eshe had to pass in order to reach the far side of the room (see Fig. 5.1). Although she had already mastered her routine, Eshe's job was far from trivial. First, she had to place her nose into a hole in the wall located just across from the aperture. Then, she had to estimate, in a single attempt and as quickly as possible, the aperture's diameter. To make things more complicated and interesting, this diameter varied randomly from trial to trial. Thus, in order to receive the reward she craved so much, Eshe had to determine whether the present aperture's diameter was narrower or wider than the one she had explored just a few seconds earlier. All in total darkness.

Without being able to see the bars, Eshe had only one way to achieve her goal: she had to rely entirely on her exquisite sense of touch and the accumulated experience of performing the task again and again in the course of the past month. Amazingly, in 90 percent of the trials, Eshe could correctly decide within 150 milliseconds whether the aperture she was touching was narrower or wider than it had been in a prior trial—even when the difference in diameter was a mere couple of millimeters.

Eshe achieved her virtuoso tactile performance not by using the tips of her paws, but by touching the edges of the two bars with the tips of her facial whiskers, the prominent long hairs (also known as mystacial vibrissae) that sprout from both sides of a rat's face. Any man trying to solve a similar task by rubbing his mustache or beard on the same aperture would fail miserably. (That would go for a woman making the attempt, too.)

Humans, of course, use their fingertips in problems involving tactile discrimination. This works very well because the skin covering our fingertips contains a very high density of mechanoreceptors, whose elabo-

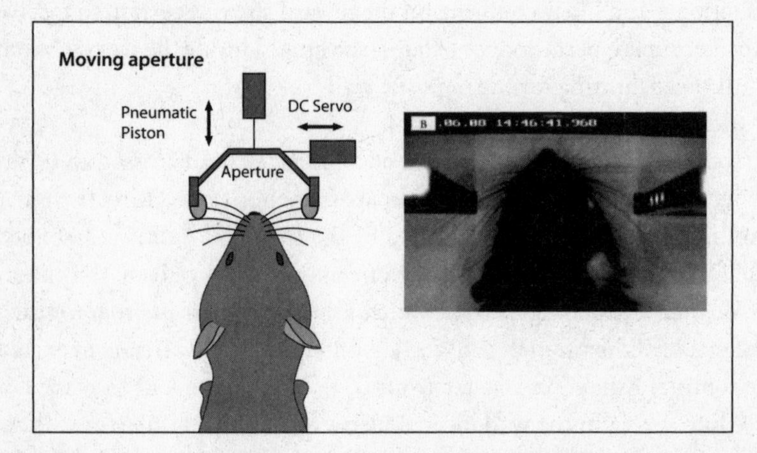

FIGURE 5.1 An experimental setup designed for testing the ability of freely behaving rats to use their facial whiskers to discriminate the diameter of an aperture in the dark. Eshe is seen in the right frame performing the task with gusto! *(Originally published in D. J. Krupa, M. C. Wiest, M. Laubach, and M.A.L. Nicolelis, "Layer Specific Somatosensory Cortical Activation During Active Tactile Discrimination,"* Science *304 [2004]: 1989–92.)*

rate and diverse morphological structure allows us to perceive minute forces applied to the body's surface. When the receptive field of each of our mechanoreceptors is activated by external forces, the information contained in a tactile stimulus is translated into the electrical language of the brain. Thus, the tactile or somatic receptive field defines the amount of skin that, when stimulated, leads a peripheral mechanoreceptor or a neuron in the central nervous system to respond by producing a salvo of action potentials. The process of relaying action potentials from mechanoreceptors, known as sensory transduction, ensures that as a mechanical stimulus is applied to the skin of a fingertip, swift sequences of action potentials are produced to signal the location, intensity, and duration of the stimulus. Mechanoreceptors generate tactile images of the world that exists immediately around us.

As we learned earlier, tactile messages are then conveyed through peripheral nerves to the central nervous system for further processing. The ascending bundles or nerve fibers that climb up first toward the subcortical relays and then the cortical areas are usually referred to as "feedforward somatosensory pathways." These feedforward pathways are matched by so-called feedback nerve projections, which flow in the opposite direction; they originate in the somatosensory cortex and project down to several subcortical structures with a variety of strange names, such as the thalamus and brain-stem nuclei. All of the sensory systems contain a similar arrangement of feedforward and feedback projections. Without knowing it, Eshe had been participating in an experiment designed to address one of the very basic functions of the interplay between these feedforward and feedback pathways. Originally conceived by David Krupa, a neurophysiologist working in my lab at Duke, Eshe's tactile discrimination task had been created to investigate the interactions in this somatosensory brain circuitry during the active exploration of objects.

That Eshe decided to use her facial hair to solve this task was therefore only proper. After all, Eshe was a rat. And when rats really need to escape from cats, running through an aperture of unfamiliar location and diameter—say, a hole in the wall of the attic—the rhythmic movements of their facial hair offer the best hope for success, as frustrated cats would surely tell us, if they could.

Like the fingertips of primates, the hair follicle of each facial whisker of a rodent contains a high density of mechanoreceptors, which translate any minute mechanical deflection of a hair into electrical signals for transmission to the feedforward nerve pathways and eventually the central nervous system. This task is performed by the trigeminal system, a subdivision of the somatosensory system specialized in conveying and processing tactile signals from the face.

Understanding how rats use their whiskers to perform their tricks of evasion is an interesting scientific question in its own right. However, many neurophysiologists realized early on that addressing the basic question of how large populations of trigeminal neurons process tactile information carries much more weight than simply learning how anxious rats elude hungry cats. Indeed, since the early 1970s, the rodent trigeminal system has become one of the favorite experimental models of neurophysiologists interested in researching neural coding. That was exactly what motivated Janaina Pantoja, a graduate student working in my lab, to spend her days "listening" to neurons from the trigeminal system. Such "brain listening" was made possible by a neurophysiological technique designed and implemented during my days as a postdoctoral fellow in John Chapin's laboratory and later expanded in my own laboratory at Duke.

In its current incarnation, the technique for chronic, multisite, multielectrode recordings enables neuroscientists to continuously and simultaneously monitor the electrical activity produced by up to five hundred single neurons, located at multiple interconnected brain structures that define a neural circuit, for periods ranging from days to many years. Over the last twenty years, the spatial and temporal sampling ranges of this method have provided an unparalleled tool for probing the brain, compared with other major technologies, as can be seen in an illustration created by Terry Sejnowski, a computational neuroscientist at the Salk Institute in San Diego (see Fig. 5.2). A generation of scientists at Duke and elsewhere has now used this method to record the concurrent activity of hundreds of single neurons, distributed across many of the brain structures that define the trigeminal system, while rats performed a variety of behavioral tasks.

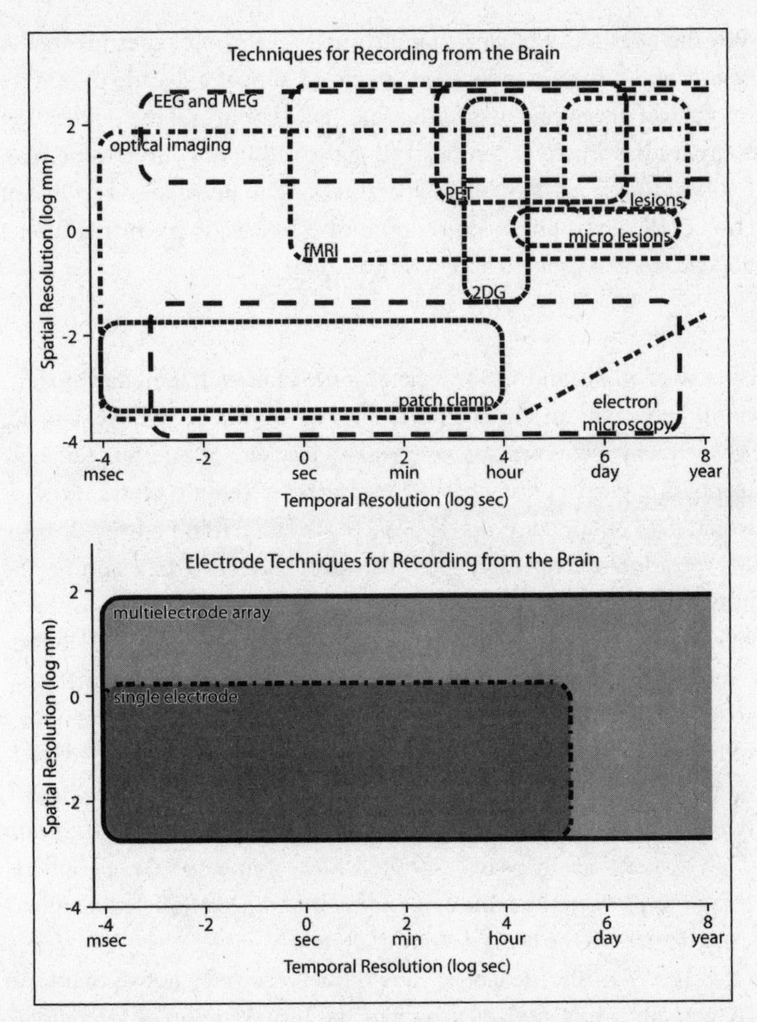

FIGURE 5.2 Temporal and spatial resolution of the single electrode versus multi-electrode recording methods. The top graph relates the spatial and temporal resolution of most techniques used to investigate brain function. The lower graph compares the single and multi-electrode recording method using the same parameters. *(Adapted from A. Grinvald and R. Hildesheim, "VSDI: A New Era in Functional Imaging of Cortical Dynamics."* Nature Reviews Neuroscience *5 [2004]: 874–85, with permission from Macmillan Publishers Ltd.)*

But the possibility of carrying out such rewarding experiments did not come about instantaneously. It required almost a decade of intense technological development, arduous data collection, and the publication of many studies aimed at demonstrating the validity of our new method. Actually, it took a *lot* of publishing and talking to persuade a significant fraction of the neurophysiology community, so used to recording from a single neuron at a time, to accept our findings.

This persuasion began in 1989, when I joined John Chapin's lab at Hahnemann University in Philadelphia, the City of Brotherly Love. During the subsequent five years, my goal was to implement and test this new neurophysiological approach systematically. A sample of the type of neuronal data obtained in this approach can be seen in Figure 5.3.

My postdoctoral research had also been designed to examine the validity of a neural coding scheme favored by most somatosensory neurophysiologists working with rodents at the time. Known as the labeled-line model, this neural coding scheme was a variation on functional localization. It posited that sensory information generated at the rat body's periphery was conveyed, through multiple parallel and segregated feed-forward somatosensory pathways, all the way to the neocortex. The model, therefore, purported that sensory information was processed by the brain through a strict feedforward circuit, which connected the peripheral mechanoreceptors in the skin surrounding the whisker follicles to higher-order structures in the central nervous system.

In the early 1970s, the labeled-line model received a boost, thanks to the American and Dutch neuroscientists Tom Woolsey and Hendrik Van der Loos, then at Johns Hopkins University School of Medicine. In a seminal study, they extracted blocks of tissue containing the entire mouse primary somatosensory cortex (S1), flattened the blocks, and cut thin tangential tissue sections, spanning the entire depth of the cortical tissue. Next, they employed a histological staining method to reveal the existence of a particular spatial distribution of cortical neurons that contain high levels of a mitochondrial enzyme—cytochrome oxidase, or CO—in the S1 cortex.

As in other mammals, the mouse cortex depth can be divided into

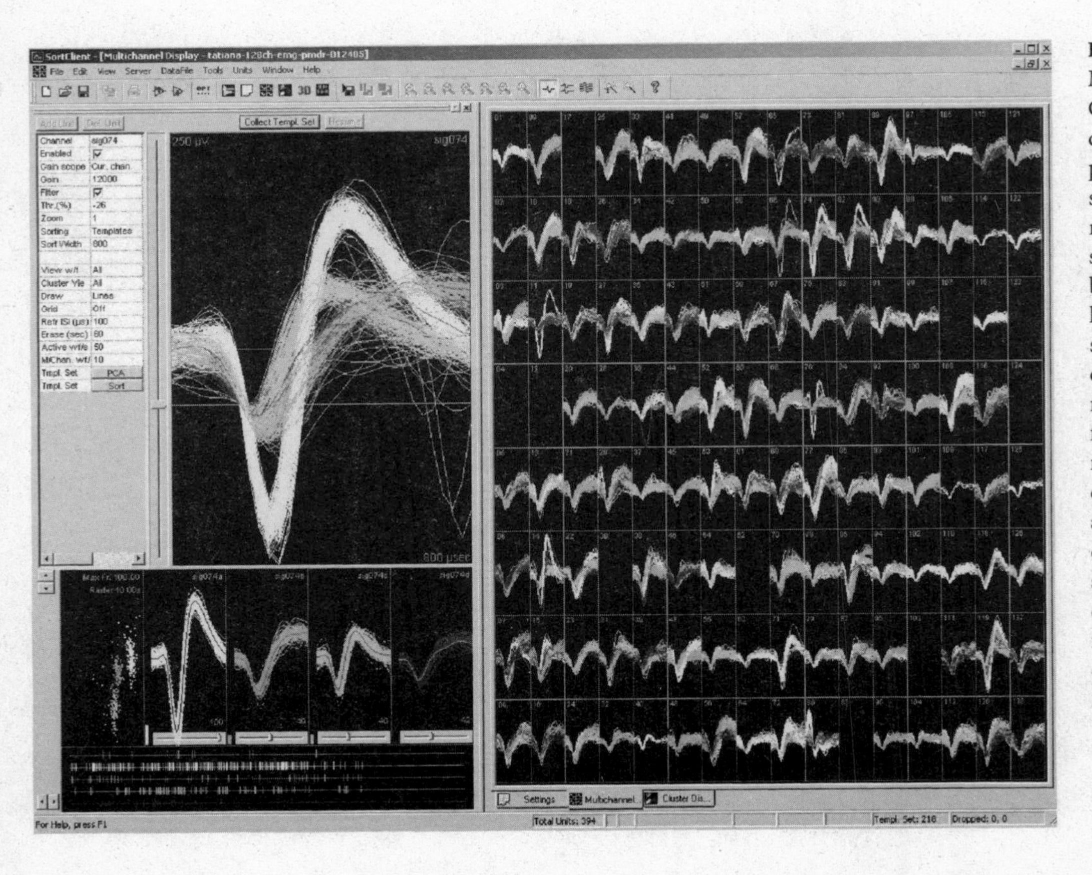

FIGURE 5.3
Many single neurons. Computer screen image depicting the action potentials produced by a sample of 394 cortical neurons recorded simultaneously in a freely behaving primate. The leftmost half of the picture shows four distinct families of action potentials, recorded simultaneously from a single microelectrode of an array, demonstrating the electrical activity of four different cortical neurons sampled simultaneously. These four distinct neurons are shown in isolation in the lower left corner. *(Courtesy of Dr. Miguel Nicolelis.)*

six layers, numbered I to VI. After staining the tissue, Woolsey and Van der Loos analyzed their thin brownish S1 sections sequentially from the top (layer I) to the bottom (layer VI). At midcortical depth (i.e., layer IV), they were surprised to discover the presence of multiple, clearly identifiable clusters of CO-rich neurons, forming a well-delineated matrix arrangement in which both rows and columns were clearly visible. Woolsey and Van der Loos named each of these CO-rich neuronal clusters a "barrel," and the entire matrix arrangement the "barrel fields." To everyone's astonishment, these barrel fields defined a beautiful, if slightly distorted, map of the entire mouse facial vibrissa. In this topographic map, each of the barrels identified the position of a particular, single facial whisker; the observed rows and columns of the barrel fields precisely matched the rows and columns that characterized the spatial distribution of whiskers on the mouse snout. Whisker rows run through the vertical axis of the face, from the topmost (A) to the bottommost (E) row, while whisker arches run through the horizontal axis, from the tail-most (1) to the snout-most whisker (5 to 10, depending on the row). Thus, each facial whisker could be identified by its row and arc position in the neuronal cluster. For instance, whisker C2 (a pet of many research projects) is the second whisker in the third row.

Woolsey and Van der Loos's map triggered a flurry of scientific interest in the rodent trigeminal system. Soon, a similar barrel-field arrangement was found in the rat (see Fig. 5.4). And it wasn't limited to S1. Topographic maps existed in the main thalamic relay nucleus of the rat trigeminal system, the ventral posterior medial (VPM) nucleus, as well as the main subdivisions of the trigeminal brain-stem complex. In the VPM, these CO-rich clusters were named "barreloids," while in the brain stem they became known as "barrelets." Overall, these histological studies uncovered stacks of topographic whisker maps at each of the subcortical relays of the trigeminal system, linking the peripheral organs, including the facial whiskers, to the S1 cortex. Further experiments revealed that neurons located within a given VPM barreloid—let's say the one representing whisker C2—tended to project primarily, though not exclusively, to the core of the cortical barrel representing the same whisker in layer IV of the S1 cortex. This supported the notion that the rodent somatosensory system represented the quintessential example of

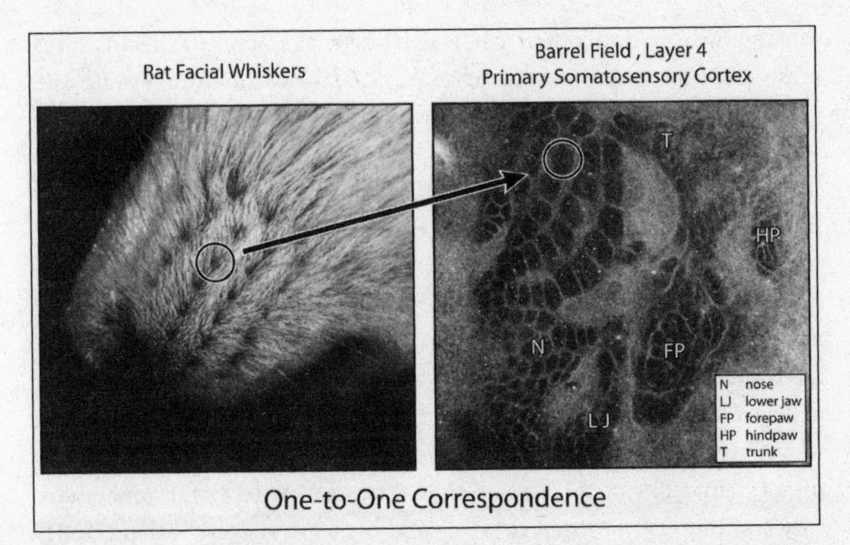

FIGURE 5.4 The whisker map of the rat's face. The left panel depicts the distribution of facial whiskers in the rat snout on four rows and many columns. The right half of the figure depicts a horizontal section through layer IV of the rat primary somatosensory cortex (S1) that contains the entire "rattunculus," including a whisker representation (barrel cortex), nose (N), lower jaw (LJ), forepaw (FP), and hindpaw (HP), and is stained for the presence of a mitochondrial enzyme found in neurons. Dark clusters represent clusters of neurons in layer IV. Notice that the barrel cortex contains an isomorphic representation of whisker rows and columns. Circles identify whisker C2 both in the rat face and the S1 cortex.
(Courtesy of Drs. John Chapin and Rick Lin.)

a labeled-line system. In such a system, the single most important prediction derived from such a neural coding scheme was that individual neurons located within each of the cortical barrels, thalamic barreloids, and trigeminal barrelets should respond significantly only to stimulation of the individual whisker that each of them represented in the overall map. That whisker became known as the "principal whisker."

Sure enough, initial measurements obtained in deeply anesthetized rats lent further support to the labeled-line model by showing that single neurons located within a given cortical barrel responded strongly, by producing a brief sequence of action potentials, to the mechanical displacement of the facial vibrissa represented by that CO-rich cluster of neurons. Over the next decade, single neuron recordings, all obtained separately from individual cortical barrels, thalamic barreloids, and brain-stem barrelets, seemed to move the labeled-line model into the realm of established scientific theory.

By the late 1980s, however, a few breaches started to appear. Leading the charge was the British neurophysiologist Michael Armstrong-James, then at University College London, who decided to record signals from single neurons located in *multiple* cortical barrels of anesthetized rats. Although Armstrong-James was able to identify the principal whisker of most of these cortical neurons and demonstrate that it corresponded to the barrel in which the neuron was located, he learned that these same neurons were also capable of responding to the mechanical deflection of whiskers surrounding the principal one. In what was considered at the time an almost heretical conclusion in the small but feisty rat somatosensory community, Armstrong-James and his team suggested that the RF of neurons in the rat barrel cortex was not confined to a single principal whisker. Instead, they said a few "surrounding whiskers" also drove neurons to produce weaker and slower but nonetheless significant responses.

It was in the summer of 1991, in the middle of this tumultuous state of affairs, that John Chapin and I decided we were ready to apply our technique for multi-electrode recording to the question of the principal whisker. We had spent two long years testing circuit boards and building microelectrode arrays. We decided to measure the RFs of individual neurons located in the many barreloids of the rat VPM nucleus of the thalamus, the main source of ascending somatosensory thalamic fibers to the S1 barrel field. The rat VPM was chosen purposely because it contains only one class of cells, named thalamocortical (TC) neurons. TC neurons have extensive and elaborated dendritic trees that receive many hundreds of synaptic contacts from ascending nerve fibers originating in the trigeminal brain-stem nuclei. In addition to these gorgeous and busy dendrites, TC neurons have long axons that exit the thalamus and project all the way to the barrel fields of the rat S1, where they form excitatory synapses with the dendrites of cortical neurons. This thalamocortical pathway completes the last feedforward section of the major neural highway that connects the array of whiskers on the rat's face to the rat's cortex (see Fig. 5.5, left panel).

There is another component to this feedforward circuit, though, that is worth mentioning. Before reaching the S1, the axons of TC neurons

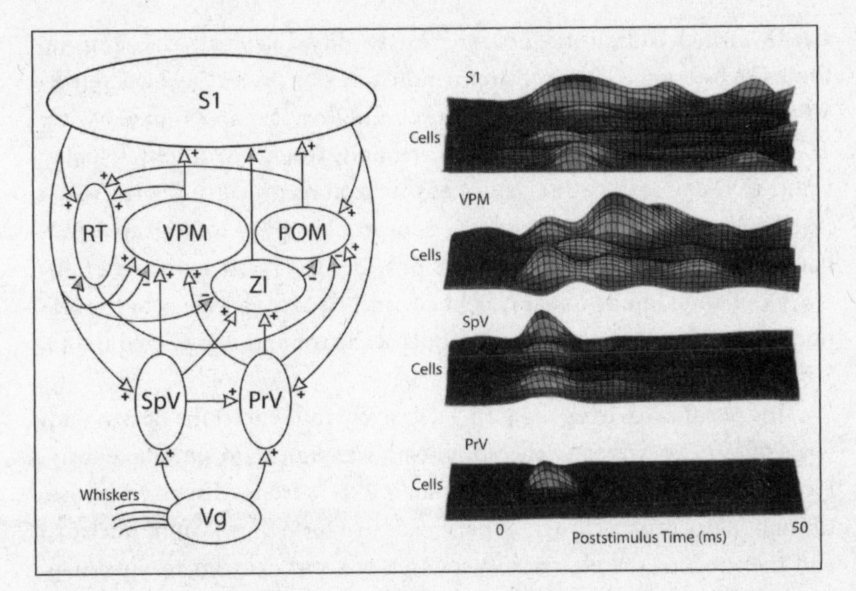

FIGURE 5.5 The left half of the figure depicts the connectivity of some of the main brain structures that define the rat trigeminal somatosensory system. Excitatory (+) and inhibitory (-) connections are shown. The mechanical stimulation of facial whiskers triggers electrical responses of neurons in the trigeminal ganglion (Vg). Vg neurons project to two distinct trigeminal nuclei in the brain stem: the spinal (SpV) and principal (PrV) nuclei. These two send nerve pathways to three thalamic nuclei: the ventroposterior medial nucleus (VPM), the posterior medial (POM) nucleus, and the zona incerta (ZI). The thalamic reticular nucleus (RT) provides inhibition to the VPM and POM. VPM, POM, and ZI provide thalamic nerve fibers to the primary somatosensory cortex. Of those, the ZI is the only one that sends inhibitory afferents to the S1 cortex. In the right half of the figure, a stack of 3-D graphs illustrates the simultaneously recorded tactile-evoked responses of populations of individual neurons at different levels of the trigeminal system. *(Adapted from M.A.L. Nicolelis, L. A. Baccala, R.C.S. Lin, and J. K. Chapin, "Sensorimotor Encoding by Synchronous Neural Ensemble Activity at Multiple Levels of the Somatosensory System."* Science *268 [1995]: 1353–58; and from M.A.L. Nicolelis, A. A. Ghazanfar, B. Faggin, S. Votaw, and L.M.D. Oliveira, "Reconstructing the Engram: Simultaneous, Multisite, Many Single Neuron Recordings."* Neuron *18 [1997]: 529–37, with permission from Elsevier.)*

branch and give rise to an axonal collateral that terminates in a shell-like, thin layer of neurons known as the reticular (RT) nucleus, which encases, like an onion skin, most of the neuronal mass that defines the thalamus. The RT nucleus contains only neurons that primarily utilize the neurotransmitter gamma-aminobutyric acid (more familiarly called

GABA), which inhibits the neurons' excitability. Curiously, the axons of these GABAergic RT neurons return the favor by projecting back to the VPM nucleus and providing the only inhibitory synapses linked with TC neurons in the VPM. This comes in handy when a neurophysiologist wants to test a new method, since any neuron recorded in the rat VPM is, by definition, an excitatory TC neuron belonging to a given VPM barreloid. Each of these TC neurons provides the main source of excitatory thalamic inputs to equivalent cortical barrels as well as to the RT nucleus, whose neurons account for all of the inhibition detected in the VPM thalamus.

This peculiar "wiring diagram" decisively influenced the design and goals of our experiments. The initial plan was simple: to simultaneously record the activity of about two dozen VPM neurons, dispersed across the multiple neuronal clusters or barreloids that form the thalamic nucleus, and then record the electrical responses of these neurons to repetitive mechanical stimulation, conducted in a sequential and random order, of most of the facial whiskers on a bunch of lightly anesthetized rats. To ensure that we could simultaneously document the electrical activity of multiple VPM TC neurons, we built custom microwire arrays and bundles that could yield nice, robust recordings beginning a week after their careful surgical implantation. Careful and slow, I should say.

It took a while, but after a few failed implantations, we realized that to obtain the best results, we had to implant the microwires in the rat brain very slowly, which helped guarantee that we did not damage the tissue. As we perfected the procedure, we slowed down a lot: to one hundred micrometer-per-minute maneuvers into the volume of the VPM nucleus, interleaved with one-to-three-minute resting periods between each step to give the tissue time to adjust to each of our micropenetrations. It was around this time that, while listening to Philadelphia's radio stations day and night, I became closely acquainted with two local icons: the Philadelphia Philharmonic Orchestra and the Philadelphia Phillies. After months of practice, with maestro Eugene Ormandy, slugger John Kruk, and relief pitcher Mitch Williams as my close collaborators, the implants started to go smoothly.

Next, it was time to wait for the rats (and me) to recover from the surgery and to test what kind of neural signal each of the implanted sensors yielded. As John Chapin had predicted in his NIH grant application, after the recovery period, when the animals were brought to the laboratory, we could readily identify the neuronal firing from most of the implanted microwires. It is difficult to describe what I felt when the fast-flowing, fluorescent green trace of an oscilloscope revealed the identity of the first action potentials recorded from those hard-conquered VPM implants. By using a specially built multichannel amplifier, we could filter, amplify, and store the electrical signals produced by every single neuron that cared to fire next to the tip of the microwires in the rat's brain. That gave us the crisp, simultaneous signals of nearly two dozen single neurons. During our recording sessions, I routinely sent the amplifier's output to a loudspeaker, and, with the turn of a knob, endowed the neuronal electricity flowing in front of us with a voice so enticing, so melodic, that like Odysseus, one would have to fill one's ears with wax not to fall in love with it.

The singing of neurons, Dr. Timo-Iaria used to call it. After three years of waiting, I was finally listening to their serenade.

Although untrained ears listening to the sound of the neuronal firing coming out of our loudspeaker would refer to it as "popcorn-popping over a noisy AM station," those neurons were virtuosos of the tiny sparks of action potentials playing across the brain. Here in our small, soundproof bunker of a lab in Hahnemann's Department of Physiology and Biophysics, we were getting close to unveiling the secrets of a brain circuit in real time. At once, we launched into an intense spree of experimentation that included twelve- to sixteen-hour recording sessions. I vividly recall one occasion when, at 5 A.M., after spending an entire day and night debugging an Eclipse minicomputer, Chapin and I paused and realized that between us we were holding an implanted rat, a screwdriver, and a computer printout. The only thing I could think to say to my bewildered but deliriously happy mentor was: "At least we can both share the same divorce lawyer."

The siren song of those bursting neurons left us no alternative but to shrug off the joke and go back to our neuronal symphonies.

Having succeeded in recording the electrical signals produced by many VPM neurons simultaneously, the next obstacle we faced was creating a delicate way to deflect individual facial whiskers. This would allow us to measure quantitatively how each of the simultaneously recorded VPM neurons responded to a well-controlled mechanical stimulus. After a few days of tinkering, Chapin and I came up with a low-tech, low-cost solution that could have come off the shelves of a Radio Shack and your neighborhood drugstore. To manufacture the device, each morning in the lab I removed the cotton covering on the end of the long wooden shaft of a typical hospital-grade Q-tip. Then, using a Swiss Army knife, I carved the very tip of the Q-tip until it resembled a sharp needle. Using Krazy Glue, I cemented the cylindrical edge of the Q-tip to the flat surface of a thick steel washer, which was tightly fitted to the metal shaft of a small electrical motor. This motor was then placed in a small metal box wrapped in a copper mesh. When connected to a ground wire, the mesh allowed us to eliminate the electrical noise generated by the motor itself. By using a simple stimulator to drive the electrical motor, we could produce a very precise movement of the motor shaft—which created an equally accurate displacement of the Q-tip rod and its sharpened tip.

Once the day's apparatus was ready for service, usually by mid-afternoon, the demanding part of the experiment started. After anesthetizing a rat that had been previously implanted with our microwire array, we laid it on a small cushioned platform inside a Plexiglas recording chamber. I then connected the rat's array to the hardware responsible for amplifying, filtering, displaying, and storing the electrical neuronal activity of the VPM neurons. Finally, I was ready to stimulate about twenty of the rat's whiskers, all located on the side of the face contralateral to the hemisphere of the brain with the implanted array.

Stimulating one single whisker isn't easy. Looking through a mounted magnifying glass, I had to position the sharpened tip of the wooden rod a mere ten millimeters from a whisker's follicle root. At this point, the whisker simply had to rest on the tip, so that I could confirm its position on the array. Once this was done, I turned on the motor powering our whisker stimulator. Each whisker stimulus lasted one hundred milliseconds and produced a half-millimeter (or three-degree) upward movement in the whisker, followed by a downward movement that returned the whisker to its resting position. This was repeated 360 times at one hertz, that is, once per second, leaving a gap of nine hundred milliseconds between each consecutive whisker stimulus. After that, I'd find the next whisker to stimulate.

While the whisker stimulation progressed, the ensuing electrical activity produced from the VPM TC neurons I had been able to identify was recorded in perfect synchrony with the mechanical whisker stimulator. Because of the long period between stimulation trials (from the brain's point of view, nine hundred milliseconds is very long), we also got a chance to record the spontaneous activity of these VPM neurons. At the end of each experimental session, we could reconstruct the RFs of each of the neurons recorded, as well as the whisker map embedded in the entire VPM. Although this map had already been described, I was looking forward to creating the first topographic representation of it based on our simultaneous multi-electrode recordings.

According to the labeled-line model, there was no doubt that the RF of each of the VPM TC neurons would strictly be confined to a single, principal whisker represented by the barreloid within which each given neuron resided. Yet, the results that emerged from those lonely summer nights tweaking whiskers were far from what was expected. After eighteen months of analysis, Chapin and I were able to demonstrate that single VPM neurons were capable of responding significantly to the stimulation of many whiskers. When we quantitatively measured the number of action potentials produced by many single VPM neurons in response to the independent stimulation of many single whiskers, we saw that the evoked firing response of the neurons was significantly above their resting activity. Then, just by counting the total number of whiskers that drove each VPM to produce a statistically significant tactile response, we reached an unavoidable conclusion: VPM neurons had humongous, multiwhisker receptive fields. The prediction made by the labeled-line model was not even close. As we reported in two articles, one published in the *Proceedings of the National Academy of Sciences* in 1993 and the other in the *Journal of Neuroscience* in 1994, some of the VPM RFs were so large that they pretty much covered the entire face of the rat.

A graphical representation of what happened when we stimulated the C1, D1, D2, E1, and E2 whiskers helps to explain our findings. Figure 5.6 depicts the type of data set obtained in the experiments, plotted in a classic graph known as the peri-stimulus time histogram (PSTH), which

I used to quantify the RF size of the single VPM neuron. The histograms represent the frequency of action potentials (on the *Y*-axis) produced by a given VPM neuron around the time (on the *X*-axis) of stimulation of a given whisker, with t = 0 ms being the time at which I started stimulating the whisker and t = 100 ms being the time at which I stopped. In these histograms one can observe several interesting things. First, before the whisker stimulation started, this VPM neuron fired very few action potentials—its firing frequency prior to t = 0 ms was very low. Yet, about five milliseconds after the stimulus started, the neuron produced a vigorous excitatory electrical response that reached an instantaneous frequency of almost 50 hertz. This so-called short-latency response was very brief and decayed rapidly, likely due to the inhibitory action of the GABAergic RT neurons linked to this VPM cell. And given the reciprocal connectivity between VPM and RT neurons, that inhibition was likely *triggered* by the VPM neuron's strong firing.

Inspecting hundreds of such histograms, I discovered that the same VPM neuron responded, albeit with a different magnitude and timing, to the mechanical stimulation of many other single whiskers. Moreover, when I looked closely at the temporal dimension of these responses I noticed a rich pattern that, in two spatial dimensions, showed that the RFs of the VPM neurons also changed over time, after the onset of the stimulus. This is depicted in what I called a *spatiotemporal receptive field plot* (see Fig. 5.7A and C).

Each of these illustrations (Fig. 5.7A, C) begins in three dimensions, with the distribution of whiskers on the rat's face represented by its row and column spatial grid (the *X*- and *Y*-axes), plotted against the firing frequency generated by a single VPM cell when each of these whiskers is individually stimulated (the *Z*-axis). Then, mapping a sequence of these three-dimensional graphs as a function of post-stimulus time, I created a 4-D plot that measured, for the first time, how the structure of the RFs of each of these VPM neurons varied in sequential steps of five to ten milliseconds.

Reviewing these plots, Chapin and I recognized that something very intriguing had emerged in our experiments. By inspecting the first three-dimensional slice, which described the spatial domain of the RF of a single VPM neuron at five to ten milliseconds (Fig. 5.7A) after the

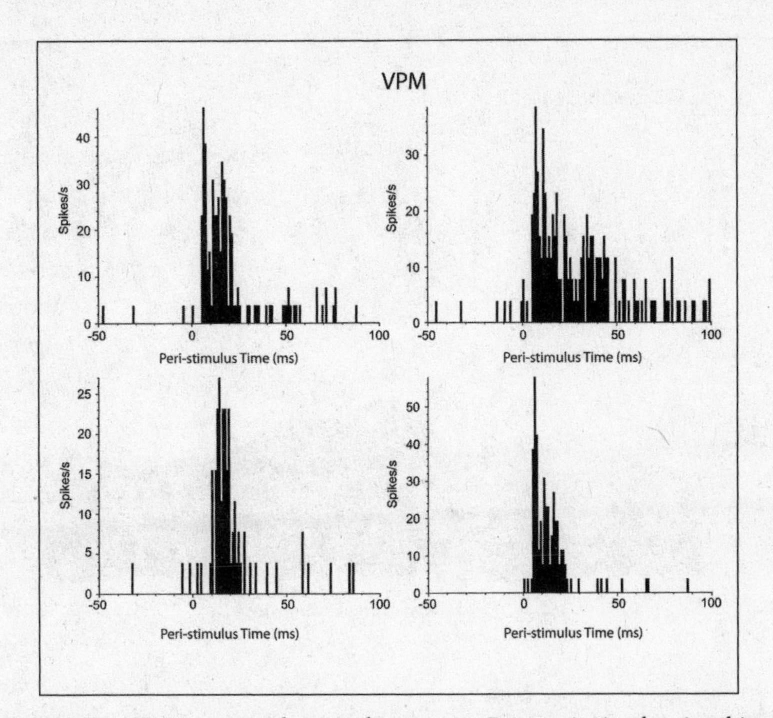

FIGURE 5.6 VPM peri-stimulus time histograms. Four peri-stimulus time histograms illustrate typical averaged electrical responses of single VPM neurons following the deflection of facial whiskers. In each histogram, the *X* axis represents the peri-stimulus time, 0 indicates the time of whisker deflection. The *Y* axis depicts the number of spikes produced by the cell. *(Courtesy of Dr. Miguel Vieira, Duke University.)*

whisker stimulus onset—the earliest possible time in which the neuron could have received information about the mechanical stimulation of the whiskers—we saw that the cell was already firing significantly. Even at this very short latency, the neuron's RF ranged across a huge spatial area, formed by somewhat weaker responses triggered by the tweaking of surrounding whiskers. Specifically for the example in Fig. 5.7A, the spatial domain of the RF was located in the farther-out, caudal region of the whisker pad (principal whisker E1), close to the curvature of the mouth where the upper and lower lips join.

Because VPM neurons responded at distinct post-stimulus times to the tweaking of different whiskers, it was also clear that the overall spatial

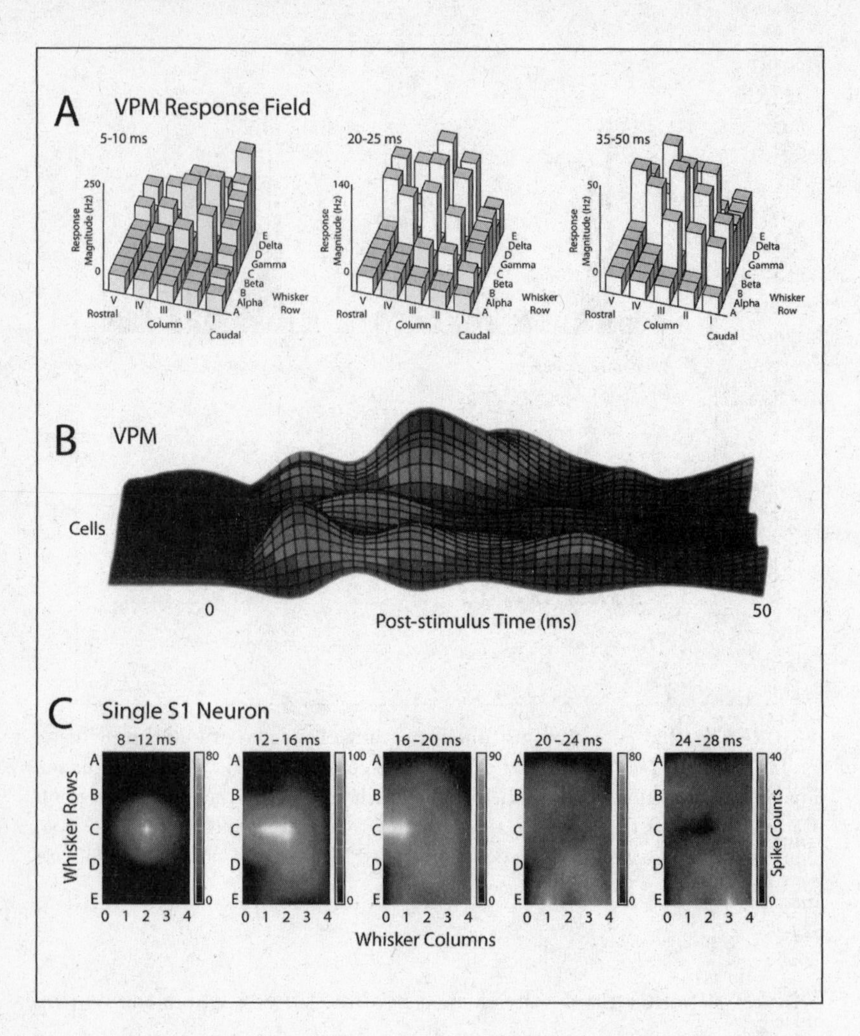

domain of each of these neuron's RFs varied significantly depending on when the stimulus happened. By twenty-five to thirty-five milliseconds after the stimulus onset, the spatial center of the RF of one VPM neuron had steadfastly migrated from whisker E1, in the caudal portion of the mouth, to whisker E4 (see Fig. 5.7A, last plot on the right), a hair located at the front of the rat snout. And the surrounding RF had been distributed, too; whereas it had started around whiskers C1, C2, D1, D2, and E2, it had moved to whiskers C3, D3, D4, and E3. Not only were the RFs

FIGURE 5.7 Spatiotemporal RF and maps. (A) Spatiotemporal receptive field (RF) of a single VPM neuron. Each 3-D graph represents the spatial domain (RF) of a single VPM neuron at a particular post-stimulus time interval (5–10 ms, 20–25 ms, 35–50 ms). In each 3-D graph the *X* and *Y* axes depict the position of whiskers in the rows and columns found in the rat's face. The *Z* axis represents the magnitude of the VPM neuron's firing response when one particular whisker is mechanically deflected. Notice that at 5–10 ms, whisker E1 elicits the strongest firing response of the VPM neuron, while stimulation of other whiskers produces somewhat smaller responses. Yet, at 35–50 ms after the whisker stimulus, whisker E4 triggers the strongest response of the same cell. Thus, the spatial center of the RF of this VPM neuron shifts as a function of post-stimulus time. (B) Spatiotemporal histogram depicting the tactile responses of a population of simultaneously recorded VPM neurons. Post-stimulus time is represented in the *X* axis, with 0 marking the onset of whisker stimulation. The *Y* axis depicts a number of individual VPM neurons recorded simultaneously. The gray-shaded *Z* axis illustrates the magnitude of firing of these VPM neurons as a function of time. (C) Spatiotemporal RF of a single neuron located in the rat primary somatosensory cortex. In each of these 3-D graphs, the *X* axis represents whisker columns, the *Y* axis represents whisker rows, and the *Z* axis (gray-scale) represents the magnitude of a single cortical neuron response. Each 3-D graph depicts a particular post-stimulus time interval (8–12, 12–16, 16–20, 20–24, 24–28 ms). Notice that, like in the VPM, the spatial domain of the RF changes as a function of post-stimulus time. *(Adapted from M.A.L. Nicolelis, L. A. Baccalá, R.C.S. Lin, and J. K. Chapin, "Sensorimotor Encoding by Synchronous Neural Ensemble Activity at Multiple Levels of the Somatosensory System."* Science *268 [1995]: 1353–58; from M.A.L. Nicolelis, A. A. Ghazanfar, B. Faggin, S. Votaw, and L.M.O. Oliveira, "Reconstructing the Engram: Simultaneous, Multisite, Many Single Neuron Recordings."* Neuron *18 [1997]: 529–37, with permission from Elsevier; from M.A.L. Nicolelis, and J. K. Chapin, "The Spatiotemporal Structure of Somatosensory Responses of Many-Neuron Ensembles in the Rat Ventral Posterior Medial Nucleus of the Thalamus."* Journal of Neuroscience *14 [1994]: 3511–32, with permission; and from A. A. Ghazanfar and M.A.L. Nicolelis, "Spatiotemporal Properties of Layer V Neurons in the Rat Primary Somatosensory Cortex."* Cerebral Cortex *9 [1999]: 348–61, with permission from Oxford Journals.)*

of these VPM neurons huge, but their spatial domains wandered capriciously across the rat's face over time.

Hidden deep in the rat brain, space and time had fused to such a level of intimate amalgamation that to speak of the probable spatial domain of a given VPM neuron's RF was meaningless, unless you also specified, in the same breath, the precise moment of post-stimulus time to which you were referring. Moreover, given that single VPM neurons did not fire the same number of action potentials during each trial of whisker stimulation, the *Z*-axis of each of the 3-D plots did not represent *absolute* firing magnitudes, but rather a simple estimate of the probability of each neuron to fire, in response to a given whisker stimulus, at a particular moment in time (see another example of the RF of S1 cortical neuron in Fig. 5.7C).

I call this the *uncertainty principle of neurophysiology*, one of ten principles that describe how a relativistic brain generates thinking from its own point of view.

THE UNCERTAINTY PRINCIPLE OF NEUROPHYSIOLOGY

One cannot define the spatial domain of a particular neuronal receptive field without specifying a particular moment in time. In other words, the spatial and temporal domains of neuronal firing are tightly coupled, defining a neuronal space-time continuum.

I propose that this space-time coupling emerges as a result of the fact that, at each moment in time, different combinations of neuronal afferent signals converge on neurons.

Our discovery of spatiotemporal RFs directly challenged the labeled-line establishment. Furthermore, instead of corroborating the notion that the VPM somatotopic map was defined by parallel feedforward neural pathways ascending from the periphery, our data suggested that such a tactile representation was the result of an asynchronous interplay of three main neural systems: the excitatory trigeminothalamic feedforward pathway, the excitatory corticothalamic feedback projection, and the potent inhibitory inputs provided by the RT neurons. These three major influences converged on different locations of the dendritic trees of the VPM neurons at distinct moments in time to define a dynamic spatiotemporal map (see Fig. 5.7B).

This *asynchronous convergence principle* is the second of the ten neurophysiological principles defining the relativistic brain.

THE ASYNCHRONOUS CONVERGENCE PRINCIPLE

The receptive field of an individual neuron and the "maps" embedded in brain regions are defined by the asynchronous spatiotemporal convergence of multiple ascending, local, and descending influences provided by a myriad of other neurons. Receptive fields and maps can only be properly defined by coupling their spatial and temporal domains in a single space-time continuum.

Taken together with the uncertainty principle, asynchronous convergence overturns the classical definition of both receptive fields and somatotopic maps, in which time plays no role. Instead, I propose that receptive fields and maps are nothing but dynamic and fluid spatiotemporal distributions of the potential probabilistic neuronal population firing patterns.

We soon realized that there was yet another major aftershock coming out of our experiments: that a dynamic VPM nucleus map should have the potential to endow VPM neurons with the ability to reorganize or remap their tactile responses quickly—indeed, immediately after any manipulation that changed the ascending flow of tactile information generated at the whisker pad. Our prediction could be tested directly by anesthetizing small patches of skin on the rat face, and then measuring the effects of this peripheral blocking on the RF of the VPM neurons. So we quickly performed a second series of experiments in the same animals.

Lo and behold, a couple of seconds after a patch of facial skin was anesthetized with lidocaine, a local anesthetic, a widespread functional reorganization of the spatiotemporal RFs of VPM neurons was triggered. As a consequence, the entire whisker map embedded in that nucleus reorganized itself to a new equilibrium point, and it did so almost instantaneously. Once again, by recording many individual neurons simultaneously across the whole VPM nucleus, we documented in exquisite detail, from the tips of our microwires, how the somatotopic map of the rat face graciously shifted to reflect the new reality of the animal's periphery.

For some systems neurophysiologists, these results, published in *Nature* in 1993, were even more shocking than our initial study. In the early 1990s, very few believed that a subcortical structure like the VPM thalamus could exhibit the kind of adult plastic reorganization extensively documented at the cortical level. A couple of years later, however, we were able to observe this dynamic and distributed view of brain function throughout the entire trigeminal system. Using yet another new strategy, I had decided to implant multiple arrays of microwires in the brains of the rats, eventually succeeding in taking recordings from up to forty-eight single neurons in a session. The neurons were distributed across the main structures that form the rat trigeminal system, including the trigeminal ganglion, two trigeminal brain-stem nuclei, the VPM, and the S1 cortex (see the example in Fig. 5.5, right panel). It was the first time in the history of systems neurophysiology that a neuronal sample from a whole neural circuit had been visualized and measured in a freely behaving mammal. Yes, you read it right: freely behaving—or, in this case, freely awake and *whisking.*

The moment had come to take recordings from awake animals and demonstrate that single whisker deflections triggered complex spatiotemporal waves of electrical activity, which spread across multiple CO-rich clusters within each of the neural structures we were simultaneously monitoring. The effect was most prominent at the level of the VPM and S1, but could also be observed in one of the subdivisions of the trigeminal brain-stem complex, meaning that individual neurons located in most of the relays of the rat trigeminal system (with the exception of the trigeminal ganglion and the main nucleus of the trigeminal complex) responded to the stimulation of multiple individual facial whiskers. At long last, we had come into contact with a distributed representation, a population neural code, playing its neural symphonies right before our eyes and ears.

Nothing could have been farther removed from the labeled-line model. Instead of highly specialized neurons (like the infamous grandmother neuron) that fire in response to a single stimulus attribute (e.g., the face of one's grandmother), distributed neural representations are formed by broadly tuned neurons, which convey small amounts of information. As such, any individual neuron's instantaneous firing activity, when taken in isolation, is incapable of either discriminating between multiple stimuli or sustaining any behavior. However, when large populations of broadly tuned neurons are working together, precise computations can be achieved. For instance, in the 1980s the Greek-American neurophysiologist Apostolos Georgopoulos, then at Johns Hopkins, reported that individual neurons in the primary motor cortex of rhesus monkeys were broadly tuned to the direction of arm movements. Georgopoulos went on to demonstrate that each of these cortical neurons fired significantly, and at different magnitudes and time spans, *before* a variety of movements were made. This made it impossible to predict the direction in which a monkey's arm was about to move from the activity of one neuron. Yet, when Georgopoulos combined the activity of hundreds of neurons, he was able to obtain precise predictions of movement direction and generate accurate arm trajectories—in a single trial. A similar distributed scheme appeared to be employed in the rat trigeminal system for representing tactile stimuli. By combining the activity of large populations of single neurons that displayed broad, multiwhisker RFs, we could extract precise and meaningful information about the rat's immediate surrounding environment—just as the rat brain was doing.

But there was more to this distributed framework than large populations of broadly tuned neurons.

Since my new multisite recording strategy allowed me to sample neuronal activity across most of the rat trigeminal system in freely behaving animals, one day, out of curiosity, I decided to observe how this circuit operated when a fully awake rat simply stood in the recording chamber without receiving any mechanical stimulation to its whiskers. In principle, it was a calibration experiment, a way to test the recording setup for the demanding whisker stimulation routine I would be performing on a perfectly immobile rat later in the day.

A few minutes into the recording session, I realized that the neuronal signals I was hearing had nothing to do with a passive brain participating in a routine calibration. That rat brain was certainly engaged in *something.* As the rat stopped moving around the chamber and attentively stood its ground, a very rhythmic sound poured out of the lab's loudspeaker. Paying close attention to the rat's state of "attentive immobility," I switched the neuronal amplifier signal to other cortical, thalamic,

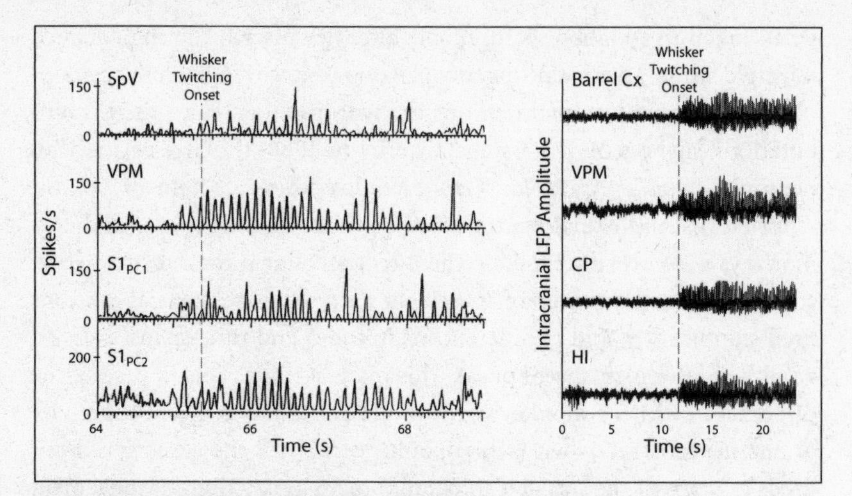

FIGURE 5.8 Examples of 7–14 Hz rhythmic mu oscillations observed in the rat trigeminal somatosensory system. In the left panel, different traces obtained simultaneously illustrate that mu oscillations start in the S1 cortex, spread to the VPM, and later to the spinal complex of the trigeminal brain-stem complex (SPV) before whisker twitching movements begin. In the right panel, a similar illustration of the relationship between mu rhythm and whisker twitching is made between the barrel cortex (rat S1 whisker area), the VPM, the basal ganglia (CP), and the hippocampus (HI). *(Originally published in M.A.L. Nicolelis, L. A. Baccala, R.C.S. Lin, and J. K. Chapin, "Sensorimotor Encoding by Synchronous Neural Ensemble Activity at Multiple Levels of the Somatosensory System."* Science *268 [1995]: 1353–58.)*

and brain-stem neurons. Most of the cells I was recording, across the entire trigeminal system, were firing at the same frequency. In fact, with the exception of the neurons in the trigeminal ganglion and in one nucleus of the trigeminal brain-stem complex, most of the cortical and subcortical structures in the trigeminal somatosensory system were expressing the same pattern of rhythmic firing.

After a few seconds, the rat, totally oblivious to my neuronal eavesdropping, delicately—and synchronously—moved the long whiskers on both sides of its face. During each movement cycle, the whiskers rapidly moved forward, then, a few tens of milliseconds later, retracted back to their original position. A new whisking cycle would begin. The overall amplitude of these whisker excursions was very small, suggesting that vibrissa behavior was distinct from the large-amplitude whisker movements used by rats to explore the objects they encounter when walking around. The most conspicuous feature of these small-amplitude whisker movements, however, was their frequency: around ten cycles per second, the same frequency of brain oscillatory activity that had preceded them.

Once they started, these discrete, low-amplitude whisker movements, which I call whisker twitches, somewhat modulated the rhythmic neuronal firing of the entire trigeminal system. As long as the rat remained in an immobile posture, both the brain oscillations and the whisker twitching could proceed uninterrupted (see Fig. 5.8). When the rat finally decided to explore the recording chamber, its whisker movements increased dramatically in amplitude, up to four to six hertz, half the frequency observed during the whisker twitching.

Having collected a couple of hours of my "calibration data," I spent several weeks poring over the neuronal recordings. To my amusement, the rhythmic neuronal firing, whose frequency power was distributed across a seven-to-twelve-hertz range, always started in some part of the S1 cortex. After spreading across most of S1, a process that took about ten to twenty milliseconds, the synchronous waves of rhythmic firing started to appear in the VPM thalamus, where almost at once most of the thalamocortical neurons were recruited into the rhythmic firing. Similar oscillations could be seen in another thalamic nucleus, and in one of the trigeminal nuclei in the brain stem (see Fig. 5.8). Even before the rat had a chance to produce its whisker-twitching movements, most of its trigeminal system was seemingly invaded by the cortex's seven-to-twelve-hertz firing wave. The firing was flowing in the opposite direction to the ascending feedforward pathways of the trigeminal system.

In a paper published in *Science* reporting these findings in 1995, Chapin and I proposed the existence of the dynamic and distributed spatiotemporal representations of tactile information that ranged across the whole of the rat trigeminal system. Moreover, we suggested that this large-scale rhythmic neuronal firing could represent an internal temporal reference signal produced by the rat brain to synchronize the activity of multiple, spatially dispersed neural structures into a cohesive circuit. This temporal signal could be responsible for generating a highly attentive state that allowed rats to anticipate and, most likely, better discriminate any incoming tactile information that could be acquired by rhythmic whisker movements before they started their next bout of active exploration.

That was my first encounter with a typical manifestation of a living brain's own point of view. By listening to those enchanting and ever-changing rhythmic symphonies of neuronal firing over our loudspeakers, I had arrived, almost by accident, at an unexplored territory of brain research.

I was eager to plunge into this new world and discover how far I could push its boundaries.

Those were the last experiments I carried out as a postdoctoral fellow at Hahnemann University. Still not quite knowing how rats escape from cats, in the fall of 1994 I found myself setting up a laboratory in the recently created Department of Neurobiology at Duke University. Soon after my arrival at Duke, a young Pakistani-American graduate student, Asif Ghazanfar, who had just graduated with a double major in philosophy and biology from the University of Idaho at Moscow, joined my research group. The Brazilian-Pakistani core of our newly inaugurated lab became kindly known in the department as the "lab from nowhere."

For the next two years, Ghazanfar and I worked furiously to test the ideas that had erupted as a result of my research at Hahnemann. For instance, Ghazanfar, who is now an associate professor at Princeton University, confirmed that single neurons in the rat S1 also exhibited large, multiwhisker dynamic RFs in which the spatial dimension of the RF varied as a function of post-stimulus time (see Fig. 5.7C). Moving further, he demonstrated in quantitative terms that populations of neurons with large, multiwhisker RFs could accurately predict the location of a single whisker stimulus, in a single trial. This was done by inputting the activity of many cortical neurons, obtained during stimulation of multiple single whiskers, into a series of pattern recognition computational algorithms, known as artificial neural networks (ANNs). In these experiments, Ghazanfar trained an ANN to employ the spatiotemporal firing patterns produced by populations of cortical neurons to classify correctly the location of a single whisker stimulus. Once the algorithm reached a high level of accuracy in classifying this "training set," he introduced a new database of trials that had never been presented to the ANNs. When the activity of populations of single neurons was fed to the ANNs, the identity of the stimulus (i.e., which whisker had been deflected) was accurately predicted, but when the activity of single neurons in isolation was fed into them, the ANNs failed.

By then, other laboratories were obtaining data from a variety of experimental methods that strongly supported our electrophysiological findings. For example, the Israeli neurophysiologist Ron Frostig, at the University of California, Irvine, employed a brain imaging method called intrinsic optical imaging to measure the spread of activation in the rat S1 cortex produced by a single whisker deflection. He was able to show that a tiny stimulus induced a complex spatiotemporal response that swept across most of S1. Moreover, in vivo intracellular recordings of S1 neurons conducted by Chris Moore, Sacha Nelson, and Mriganka Sur at MIT, and independently by Barry Connors's lab at Brown University, revealed that single cortical neurons, no matter which cortical layer they belonged to, received afferent information from many whiskers. Thus, each neuron was bombarded by synaptic currents that were triggered by the stimulation of many single whiskers. As in our results, the RFs of these neurons were multiwhisker.

Ghazanfar went on to show that VPM and S1 neurons could also integrate mechanical stimuli delivered to multiple whiskers at once, and that blocking S1 neuronal activity decisively affected the multiwhisker tactile responses of VPM neurons. A few months later, David Krupa, then a postdoctoral fellow in my lab, observed that blocking the so-called corticothalamic pathways decreased the ability of VPM neurons to exhibit plastic reorganization following anesthesia of a few facial whiskers. These findings gave credence to the asynchronous convergence theory by demonstrating, unequivocally, that the feedback somatosensory projections, originating in the S1 cortex, that target the VPM thalamus sometimes play a major role in managing the flow of tactile information through the thalamus. Based on these various results, we proposed that the highly dynamic, multiwhisker tactile responses of both S1 and VPM neurons were determined by the asynchronous convergence of a multitude of ascending, descending, local, and modulatory afferents that converge to each of these neurons at a different moment in time.

Many of the predictions that could be derived from the asynchronous convergence principle required extensive testing. For instance, while a graduate student in my lab in the late 1990s, Erika Fanselow, now at the University of Pittsburgh, designed a clever way to measure how S1 and VPM neurons in freely behaving rats would respond to similar tactile stimuli delivered under different behavioral conditions. By implanting a tiny "cuff" electrode around the infraorbital nerve (ION), the branch of the trigeminal nerve that innervates the facial whiskers, Fanselow delivered precise sequences of electrical pulses to the ION while simultaneously recording the evoked responses of populations of single neurons. She then employed this apparatus to measure how these neuronal responses varied according to the typical, and stereotypical, behaviors exhibited by rats during their daily cycle.

When rats were moving their whiskers, their cortical and thalamic neurons responded to tactile stimuli in a way that was very different from the responses observed when the same animals were quiet and immobile. Instead of the classical cycles of excitatory responses followed by profound, long-lasting inhibition, the cortical and thalamic neurons of the rats responded in a more sustained manner to a single electrical nerve pulse and did not exhibit any post-excitatory inhibition—no matter what sort of whisker movement they were producing. This prompted Fanselow to deliver to the nerve a sequence of two electrical pulses instead of one pulse. What she noticed was astounding. When rats were awake, immobile, and not producing any whisker movements, their cortical and thalamic neurons could only respond to the first electrical pulse; the second pulse was masked by the neurons' post-excitatory inhibition. On the other hand, when rats were actively moving their whiskers, their S1 and VPM neurons could respond well to both electrical pulses; they could even respond to pulses separated by as little as twenty-five milliseconds. Obviously, whisking allowed both the cortex and the thalamus to represent faithfully a sequence of tactile stimuli, something that was not possible when the rat was simply awake but immobile.

Fanselow's results clearly showed that tactile responses varied depending on the animal's behavior state. Of course, her subjects were not engaged in a *meaningful* tactile task. So the question arose: how would the somatosensory system of the rat behave when it needed to use its whiskers to perform a meaningful and demanding task, such as using its facial hair to judge the ever-changing diameter of a hole? As every household cat knows, this is a task rats perform with gusto.

While David Krupa worked on designing an appropriate experimental task to address this issue, another member of our research team, Marshall Shuler, now an assistant professor at Johns Hopkins, discovered that a large percentage of S1 neurons outside the barrels of layer IV responded to the stimulation of whiskers located on both sides of the rat's face. These bilateral responses were first observed in lightly anesthetized animals. A few years later, Shuler's experiments were repeated with awake rats by Mike Wiest, a postdoctoral fellow in my lab at Duke, now an assistant professor at Wellesley College. He confirmed Shuler's findings: rats judged the diameter of a hole by integrating tactile information generated by whiskers on both sides of the face.

By then, Krupa had found a way to train rats to perform the task that Eshe later mastered so well. This allowed us to explore whether tactile responses in S1 and the VPM nucleus varied according to whether a multiwhisker stimulus was delivered passively to the rat or whether it resulted from the active engagement of the animal's whisker in a tactile discrimination task in which success was rewarded. To control for the possibility that the whiskers could be stimulated differently when the animal actively touched the aperture, Krupa built the ingenious apparatus in which the aperture created by two bars can move toward the face of an awake but immobilized rat. In this setting, the animal's whiskers are rubbed by the edges of the bars in a way that is virtually identical to what would happen if a rat rushed into a box and touched the bars by itself. The only difference was that in the active task, the rat had to use its whiskers to discriminate the aperture's diameter and behave a certain way to receive a liquid reward.

The experiment showed that if multiple whiskers were stimulated passively, either by a multiwhisker stimulator or by the movement of the entire apparatus toward the animal's face, both S1 and VPM neurons produced short-duration, phasic excitatory responses. Very few pure inhibitory responses were observed. However, when animals actively engaged their whiskers to judge the diameter of the aperture in exchange for a reward, a large percentage of their S1 and VPM neurons exhibited

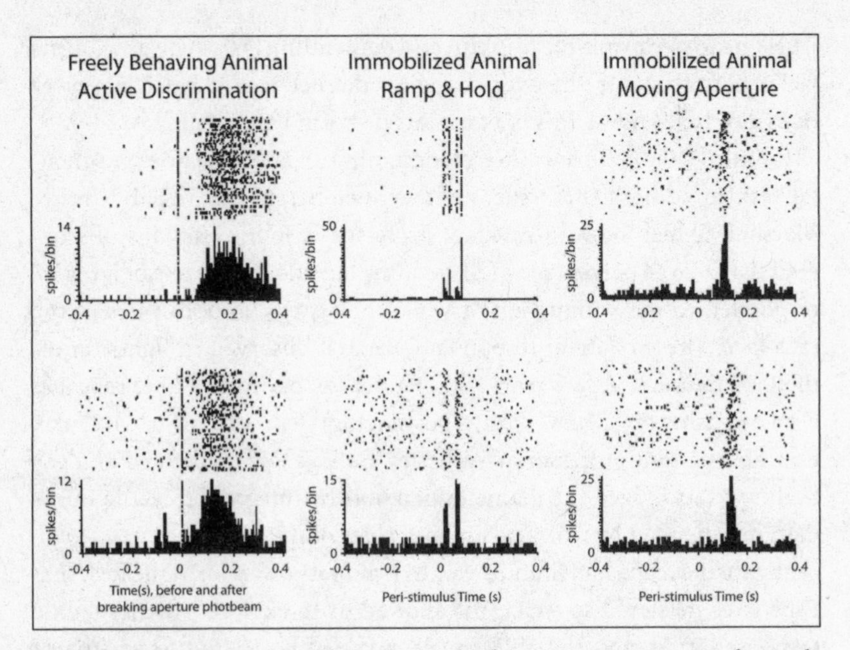

FIGURE 5.9 Peri-event histograms depict the firing response pattern of a single cortical neuron in the rat primary somatosensory cortex under three different behavioral conditions: active tactile discrimination in a freely behaving animal (leftmost panel), awake but immobilized (center panel), and immobilized and passive discrimination (rightmost panel). Notice how the pattern of this neuron's responses is totally different according to the animal's behavioral context. For each histogram, the *X* axis represents peri-event time, with 0 indicating the onset of facial whisker mechanical stimulation, and the *Y* axis represents the electrical firing response of the neuron in spikes per second. *(Originally published in D. J. Krupa, M. C. Wiest, M. Laubach, and M.A.L. Nicolelis, "Layer Specific Somatosensory Cortical Activation During Active Tactile Discrimination."* Science *304 [1989–1992, 2004].)*

intense, long-duration excitatory responses (see Fig. 5.9). Moreover, a large percentage of cortical neurons exhibited pure long-lasting inhibitory responses that had never been seen in either anesthetized or awake-but-immobile rats.

When an analysis of the responses across the different cortical layers was conducted, Krupa discovered that the neuronal firing in layers II/III and V/VI tended to increase or decrease early in the trial, long before the rat's whiskers had made contact with the bars. More surprising, modulations in the neuron firing rate started before any neuronal firing could be seen in layer IV, the main target of ascending thalamocortical fibers that carry tactile information from the periphery to S1. By using a variety of physiological measurements, Krupa went on to show that during the rat's execution of the task, neurons located in different cortical layers behaved distinctly. His finding challenged the notion, championed by the illustrious neurophysiologist Vernon Mountcastle, that neurons located across a vertical column of S1 fire similarly to the same tactile stimulus. At least in the rat somatosensory cortex, when rats use their whiskers to solve a tactile puzzle, the functional unit of thinking is not a column of neurons, but populations of neurons distributed across the entire 3-D volume of S1 cortex.

As a final demonstration, Krupa fed an artificial neural network with the spatiotemporal firing patterns generated by populations of individual neurons during the execution of this behavioral task. This allowed him to show that the combined activity of up to fifty cortical neurons, which exhibited either long-lasting excitatory or inhibitory tactile responses, could predict with great accuracy whether rats were going to correctly identify a wide versus a narrow aperture diameter, in a single trial.

Using the same aperture assessment task, Janaina Pantoja, Mike Wiest, and Eric Thomson showed that some of the anticipatory neuronal firing in layers II/III and V/VI emerges during the rat's training. Even after the animals cease to touch the bars with their whiskers, S1 neurons, and to a lesser degree VPM cells, continue to exhibit firing patterns that represent the identity of the tactile stimulus for several hundred milliseconds. In fact, this sustained activity lasts until the animal is rewarded. The spatiotemporal activity of populations of S1 neurons even provides reliable predictive information about the animal's expectation of a reward, and thus whether it is likely to identify the aperture's diameter successfully.

After a decade of tweaking whiskers and listening to the brains of rats, my research team at Duke was getting close to understanding how sneaky Jerry the mouse always escapes Tom the unlucky cat. Yet, despite all the evidence gathered in these experiments, we could barely entice any of the main proponents of the labeled-line, feedforward model of sensory processing to consider that the brain is not a passive decoder of information but a dynamic and distributed modeler of a reality comprised of a multitude of feedback, local, modulatory, and feedforward neural pathways conjuring a vast and elaborate organic spatiotemporal grid. For us, the rat somatosensory system was the epitome of a new paradigm of brain function, one in which an active, ever-changing, ever-adapting brain was always ready to express its own point of view and its expectations about the surrounding world, even before real information about this world reaches its central structures via an array of sensory channels. But outside the small and highly inbred community specializing in studying the "rat barrel cortex," very few thought that our data mounted a credible challenge to the holy canon of neurophysiology.

To prove our point, we decided to do just as Clint Eastwood had done in the 1982 thriller *Firefox*, when he managed to steal a secret plane from the Soviet Union using a helmet that allowed him to think in Russian and fly the aircraft without moving a finger. Over a Philly cheesesteak dinner, John Chapin and I decided that we, too, would link a brain to a machine and make that machine obey the brain's voluntary motor will through the force of thinking alone.

As they say in academic circles, peer-review support is the best tonic a scientist can get for his ideas. Sure enough, we immediately received ours that evening when a truck driver, who had been listening attentively to our conversation from a nearby booth, promptly gave us a thumbs-up and said: "That ain't a bad idea at all!"

6

FREEING AURORA'S BRAIN

Sitting comfortably in her favorite laboratory chair, taking a few sips of fruit juice, she seemed absolutely relaxed. For the past few weeks, she had been at the top of her game. And she made everyone around her aware of that.

No insecurity, no inferiority complex. That night, like on many recent nights, she exuded confidence. She was the quintessential go-getter, ready to make her indelible mark in the world of science.

Despite being relegated to sidekicks, we played along. Because above all, we loved Aurora's larger-than-life persona.

She had overcome disappointments, difficulties, and in some cases bare injustice—and she was quite willing to let you know it. Like other pioneers before her, she had endured these trials with a love for adventure and discovery. Tonight, she would reach the pinnacle of her scientific career.

Indeed, in a mysterious and strange way, Aurora had become one of us. A coinvestigator, decisively contributing to research that was pushing beyond the limits of our understanding of the brain. She was a member of a team assembled to demonstrate what, just a few months earlier, had seemed impossible. Yet, no one involved in the often cruel business of academic research could claim that her achievement was anything but spectacular.

Not at all.

An unproven middle-aged worker who had failed at most everything she had previously tried in life, Aurora had been forced to jump-start her career. She had proved her scientific mettle through her own hard, meticulous, and sometimes dull laboratory work. No free lunches. No big break. No free ride just because she was cute.

Which she was, by the way.

And a serious flirt too, I might add.

By a long shot Aurora had become the golden girl of the lab. Indeed, many of her male peers seemed pretty jealous of the VIP status she enjoyed. But life was not always easy for Aurora. At the end of her initial training, she had not had much to show for her labors. She had no scientific papers to gloat over, no good data to publish. Her citation index was, mildly put, unmentionable. Her funding track record was, well, next to nil. That wasn't so uncommon. But to her total dismay and dejection, she had been denied a full-time job in a prestigious federal research institution, located in the suburbs of Washington, D.C. In the terminology used by that prominent institution, she was known simply as a "Reject." In their report, they said that she was hopelessly opinionated, overambitious, and too creative for her own good.

In fairness, she had not been easy to mentor. Every experienced scientist who tried to guide her to some task that needed to be investigated quickly realized that Aurora held very high standards for what work she considered worth her attention. For no clear reason, nearly every scientific project offered to her by senior scientists was deemed totally unsuitable. If she thought your idea was bad, she wouldn't work on it.

It was a problem that Aurora was a bit narcissistic. She had done more than her share of waiting. A slow learner, she fell behind her peers' performance and often had a tough time recovering from her mistakes. While she was struggling, many of her peers (males, I should point out) were being labeled "star performers." They seemed almost disdainful of her. Some even bet, behind her back, that she would never make it.

She was so tough-minded that it came off as a display of a ferocious disregard for authority. She could be really mean, without showing a drop of remorse in her face or conscience. Occasionally, out of sheer desperation, people in her lab would beg on their knees for her to perform accordingly. Others cried like children, to no avail. She showed no mercy and even contributed to some people dropping out of science outright. Or so the Aurora legend went.

Her perseverance and principles paid off, though, and the memories of her humbling start made her current prospects even sweeter. Looking from the top of her chair, with those penetrating and rather defiant black eyes, she would mumble something, drink her fresh orange juice, maybe take a brief nap, oblivious to all and everyone, a free spirit with no respect for weaklings or traditional behavioral neuroscientists.

Above all, Aurora was about action. Once in the lab, she unleashed pure, massive amounts of adrenaline. She wanted to perform every test, every game, all the time, and super fast. Forget about absurdly complex or boringly repetitive oculomotor tasks. She wanted experiments that provoked novelty, excitement, risk, and intensity. As later events proved, even a "Reject" like Aurora, given another chance and a nurturing environment, can hope to stamp a major contribution into the big book of science.

One day, during a phone conversation with a great friend of mine, who happens to work in a reputable federal research institution, I mentioned, casually, that I was looking for a collaborator on a new project.

He immediately and enthusiastically offered to send Aurora to work for me.

That was very nice of him, I thought.

My friend, however, did not elaborate too much on Aurora's past record of achievements in his lab.

I have to admit that my first meeting with her was neither auspicious nor pleasant. She displayed a very arrogant demeanor. No camaraderie was shared between us. In fact, I came away from that meeting with the impression that although she was supposed to be a pretty reasonable worker, I had better not try to impose my way of doing things on her. Her scrutinizing stare implied that we could be colleagues, maybe friends, *if* I let her be herself.

So I did exactly that.

During her first few months in the lab, she simply abhorred befriending postdoctoral fellows, technicians, graduate students, even the cleaning staff. Anyone who tried to talk her into psychological nonsense and modern techniques of behavioral training was condemned to scorn. She was not about to give up her long-held beliefs unless the payoff was substantial, and sweet—very sweet. Aurora, as we later discovered, was really addicted to fruit juice. In particular, she would do anything you asked if a pitcher of Brazilian orange juice was set down next to her.

Then, something quite unexpected happened. Out of nowhere, one night in the fall of 2001, Aurora decided we were worth her attention and collaboration. She even began to simulate an easy half-smile around the graduate students.

True, she could still be touchy. As she was throwing one of her renowned tantrums, she once tried to scratch a coworker. Fortunately, she did not succeed. Some of us were shocked, but to be honest, most of us knew that if you messed with Aurora, particularly when she was having her lunch, you did so at your own peril.

During a severe admonishing session, in which she seemed not to pay a second of attention, I told her this type of behavior would not be tolerated. It never happened again. At least not in front of any of us. You see, Aurora could be sneaky, too.

Still, within months, Aurora had been amazingly transformed into the best performer at any task I asked of her. Night after night, she was a full member of our team in the little bunker of a lab at Duke.

My own motivation for accomplishing our research program surged with the arrival of unexpected news from Brazil: Dr. César Timo-Iaria had been diagnosed with a terrible neurological disorder, amyotrophic lateral sclerosis (ALS), a disease that would claim his life four years later.

For most of us, it is virtually impossible to imagine how terrifying it would be to lose control over our body movements, a muscle at a time, through an inexorable process of wasting that culminates with the failure of our most resistant musculature, the one that allows us to breathe. That is exactly the destiny of patients who suffer from ALS, which is best known to the public for its devastating progression in the life of the beloved Yankees slugger Lou Gehrig.

In one of those incredible ironies that only life can stage, Timo-Iaria had begun his career working on a new method to diagnose ALS. In the late 1950s, as a postdoctoral fellow in New York City, he became one of the first neurophysiologists to observe that patients suffering from ALS started to exhibit a continuous decrease in the conduction velocity of peripheral nerves. That meant that as the disease progressed, it took more and more time for an electrical impulse to travel through a nerve and activate a muscle. Forty years later, Timo-Iaria, then an emeritus professor of physiology at the University of São Paulo Medical School, calmly informed me that his diagnosis had been confirmed using a modern variation of the test he had perfected as a young scientist.

Throughout his last years, Timo-Iaria followed with great interest the experiments we were conducting at Duke. The primary reason for his interest, however, was not the possibility that he could benefit from what we were doing. Being a very experienced and accomplished neurophysiologist, he knew that there would be no time to translate our discoveries in time for him. But he was thinking of the possibilities for future patients and the impact these experiments could have in the field of neuroscience.

A couple of years before I met Aurora, John Chapin and I had decided to create a real-time platform to demonstrate that populations of neurons, rather than a single brain cell, should be considered the true functional unit of the central nervous system. Because most of our past work had been conducted in the rat somatosensory system, several of our colleagues in neurophysiology had openly challenged whether the neural coding scheme we had proposed, known as distributed coding, was actually significant when animals were sustaining meaningful behaviors, such as moving around or identifying objects in their surrounding environment. Rats, after all, do not have the habit of talking back to the scientists training them, so we had no way to verify what our rodent subjects were feeling, in terms of tactile experience, as spatiotemporal waves of electrical activation spread across their cortices when one of their facial whiskers was deflected. In fact, some skeptics argued that the spatiotemporal complexity of both the single-neuron receptive fields and the somatotopic maps we had identified in the cortical and subcortical relays of the trigeminal pathway were pretty much meaningless when it came to real rat behaviors. From a perceptual point of view, they said, the rat brain could basically ignore all this undesirable complexity simply by taking into account the strongest tactile responses of a very small number of cortical neurons to decide what type of tactile message a whisker stimulus conveyed. In this process, known as "response thresholding," an arbitrarily high activity threshold would be set by the brain to consider a given neuronal response as significant for eliciting a perceptual experience. In the rat trigeminal system, this threshold could be set so that only the high-magnitude firing and short-latency responses elicited by the stimulation of the principal whisker of a given neuron would be taken into account by the brain to build a tactile image of the external world. Accordingly, the smaller, long-latency responses of the dynamic "surrounding" components of the neuron's RFs and the maps would, rather conveniently, be filtered out and eliminated as the somatosensory system interpreted tactile sensory information.

That was one quick and sanitized way to dispense with the headaches set off by our findings.

Yet, despite finding this elegant solution for handling our "troubling" data, our colleagues were a bit more vague about how the brain would actually determine the level of this threshold and how networks of neurons would know the difference between so-called useful action potentials and those that should be filtered out. Why, we asked, should such rich complex neuronal dynamics be considered a problem, a *villain*, something that should be swiftly eliminated from the cortex (as well as from our theories)? By proposing that a response thresholding took place across the rat somatosensory system, our colleagues were in essence removing the temporal dimension—the fourth dimension—from tactile responses and somatotopic maps. It was as if a dynamic mammalian brain was too undesirable a nuisance to enter into their models of how touch sensations emerge. Only static spatial relationships could, in their view, give rise to tactile perception. In their model, the sensation of touch solely originated at the skin's epithelium and then through parallel feedforward labeled lines extended, more or less completely segregated, through the stack of multiple, distorted topographic maps all the way to layer IV of the S1 cortex, where it blossomed into a precious pattern of electrical interactions between neurons that reproduced the spatial nature of the peripheral tactile stimulus. These neuroscientists were still following the dogma established by Vernon Mountcastle in his studies in the 1950s: spatial order, not temporal chaos, was the stuff from which the sense of touch emerged. So, if you passed your fingertip over an embossed letter *A*, your cortical neurons would precisely reproduce the spatial organization of *A* inside your head. According to this view, time played no role in the brain's representations of the external world.

To prove that a completely antagonistic theory was closer to reality, Chapin and I had to demonstrate that populations of neurons, working together as part of widespread neural circuits, could encode enough information, using dynamic spatiotemporal patterns of activity, to sustain a motor behavior. We could no longer straightforwardly observe and quantify the physiological properties of the individual cells we happened to record while our freely behaving animals carried out a particular motor task. Although this was the classical way in which virtually all cortical physiologists, including those who defended the relevance of population coding schemes in the motor cortex, operated, our experimental strategy had to be reinvented. To convince our colleagues, we would have to introduce a completely new experimental paradigm to the field of motor cortical physiology, which is how we came up with the idea of the brain-machine interface (see Fig. 6.1).

In our original conception, chronic, multisite, multi-electrode recordings would be used to sample simultaneously the activity in as many cortical neurons as possible in an individual animal subject. Our chronic implants allowed us to maintain viable recordings of single neuronal activity for weeks to months, depending on the animal species utilized in a given experiment. All we needed to do was record populations of neurons, spatially distributed across multiple cortical areas in the frontal and parietal areas, while our animals learned to perform clear and easily quantified movements of their limbs in a well-controlled behavioral task.

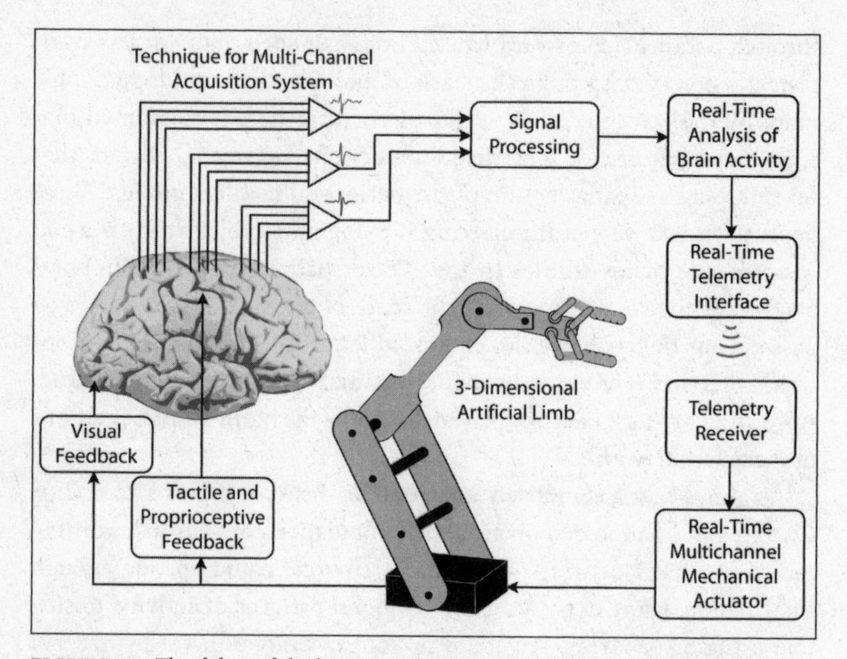

FIGURE 6.1 The debut of the brain-machine interface. Systems engineering drawing depicting the general organization of a brain-machine interface. Multi-electrode arrays and microchips are used to record large-scale brain activity. Signal processing techniques are then used to translate raw brain activity into digital commands that can be employed to reproduce, in a robotic arm, the voluntary motor intentions generated by the brain. Visual, tactile, and proprioceptive feedback signals from the robotic actuator are then sent back to the subject's brain. *(Originally published in M.A.L. Nicolelis and M. A. Lebedev, "Principles of Neural Ensemble Physiology Underlying the Operation of Brain-Machine Interfaces."* Nature Reviews Neuroscience *10 [2009]: 530–40.)*

Moreover, instead of simply measuring the physiological properties of each of these neurons, we came up with a different approach altogether. During each of our experiments, after amplifying and filtering the individual action potentials generated by the cortical neurons we were monitoring, our multichannel recording system would start to send these neuronal signals to a microcomputer. There, the continuous stream of neuronal data would be fed, as close to real time as physically possible, into a series of simple mathematical algorithms designed to optimally combine the spatiotemporal patterns of neuronal activity so that one could extract, from the overall neuronal population activity, the type of motor control programs that were normally used by the subject's brain to generate limb and hand movements for a particular task. Although there are tens of millions of neurons in a monkey's primary motor cortex alone, our technology limited us to sampling only about a hundred neurons from one cortical region at a time. That meant that our first BMI would be driven by something close to 0.000001 percent of the overall neuronal population available to the motor cortex.

For animals to operate a BMI consistently, the neuronal activity had to be translated into digital control signals very quickly—in this case, in no more than two hundred to three hundred milliseconds. This narrow timing window was not accidental. We wanted to reproduce limb movement in an artificial device under the same strict limits that governed an animal's typical reaction time, and in rats as well as monkeys it takes about that much time for the brain to generate a motor plan and enact it through the movement of the limbs. Interestingly, we would later learn that if the time for executing a BMI's operations was much longer than two hundred to three hundred milliseconds, the subject animal promptly became uncooperative and, in most cases, gave up on the experiment. Assuming the continuous outputs of our simple mathematical models carried enough information about the key motor parameters involved in the generation of limb movements, we predicted that a fully functional BMI could efficiently transfer these digital brain signals to a robotic device—for example, a prosthetic arm—and make every inch of its metal, plastic, and wire fulfill the ultimate dream of machines: to suddenly, and almost miraculously, morph into a living and feeling clump of purposeful living flesh, capable of faithfully carrying out the desires of its newly assigned master, an animal's brain.

But there was more to our BMI than simply making a limb move. To make its biological master aware of its loyal performance, the prosthetic limb would, at each moment in time, reciprocate its continuous blessing by sending sensory signals back to the brain. Given the technical difficulties at the time of accurately mimicking tactile feedback from a prosthetic limb to a rat or a monkey brain, we opted to restrict ourselves to visual feedback information delivered to the brain by including the robotic arm, or the outcome of its movements, in the animal's visual field. Thus, every time the robotic arm moved under the command of the animal's brain signals, the subject would be able to evaluate directly, through visual information, how well the BMI was operating.

The challenges facing us were enormous. First, we had to get access to enough good neurons from our chronic brain implants. Since there was no possible way to record the entire population of neurons in the primary motor cortex (M1), we had to rely on a relatively small sample of neurons to demonstrate our thesis that ensembles of neurons rather than single cells define the true functional unit of the brain. Second, our neuronal recordings had to last long enough for our animals to learn and master the motor tasks we designed to test the operation of our BMIs. Third, no one had ever described a computational algorithm robust enough to extract multiple motor commands out of raw brain activity in real time, or efficient enough to run on a budget-conscious Dell workstation, the only computer hardware we could then afford. Finally, no one knew how animals would react to the scene of having an artificial limb performing a task that they had been supposedly trained to complete using their own biological limbs.

Surprisingly, despite the reviewer's pugnacious reaction to our proposal, NIH actually funded the original BMI experiments that John Chapin and I outlined. With some seed money secured through this small research contract, by 1997 we were able to carry out the first experiments that allowed us to measure the performance of a few rats operating a true brain-machine interface. Right off the bat, our BMI design included a complete closed-control loop apparatus, which meant that the BMI utilized brain-derived signals to control the one-dimensional movements of a basic artificial device while permitting the rat to have continuous access to the device's performance through information gleaned in its visual field.

Over several weeks, Chapin and his team trained six rats, a tediously slow process that required all the patience a neurophysiologist could muster. First, the rats had to learn to press a bar with one of their fore-paws, and not with their furry butts, the preferred, if unorthodox, method usually chosen by most rodents when faced with solving this particular behavioral riddle. Once the rats had figured out how to make the forepaw movements we wanted them to use to press the bar, they then had to learn how to repeat these movements over long periods of time—persevering for several minutes per recording session, in order to give us a reasonable amount of data to feed through our computers into the BMI.

In this setup, rats pressed a bar that was electronically connected to a metal lever equipped with a tiny cup. If the rat's forepaw pressing was delicate enough, the lever moved so that the cup was positioned under a water-dripping tube. By holding the cup in that position for a second or so, the rat could use the lever to collect a drop of nice, cool Philadelphia tap water. Then, by slowly releasing the pressure applied by its forepaw on the bar, it could make the lever bring the cup all the way back to its mouth so that it could enjoy a drink. Once a rat mastered this simple motor task, a microwire array was implanted in its M1 so that we could harvest the neuronal electrical activity for the BMI. Our next goal was to get the rat to repeat the whole operation of receiving its drop of water, but now by moving the lever controlled by the BMI instead of its forepaws. To do so, the rat had to use its brain activity to control the lever's movements and deliver some refreshing water into its mouth.

It was then that we entered what could only be described as the "twilight zone" of neurophysiology, a place where the main question hovered in the air: could a rat grasp the concept of having to rely on thinking alone to get that drop of water, literally without moving a whisker?

Following a couple of weeks of postsurgical recovery, Chapin's rats returned to the experimental setup and were prepared, over a few days, for the first test of the BMI. Careful inspection of each of the brain implants revealed, to our total delight, that in each of the animals we could identify up to forty-six motor cortical neurons firing robustly during the execution of the bar-pressing task. By monitoring the simultaneous activity of these neurons while the animals moved their forelimbs to press the bar, we soon realized that most of these neurons exhibited "premovement activity"—the preparations made by the M1 cortex in the two hundred to three hundred milliseconds prior to producing a body movement. This meant that we would be able to record the high-quality brain activity necessary to make a BMI work.

Now, as the rat commanded its forelimb to press the bar, Chapin recorded the action potentials produced by the microwire array's sample of motor cortical neurons. A parallel array of electrical resistors had been assembled into an integrator board so that the contribution of each individual neuron could be appropriately weighted—and given the computer's processing power, this had to be done manually during each recording session. These weighted contributions were then summed up to generate a single, continuous analog "motor control signal" output, which would create a brain-derived signal capable of predicting the rat's voluntary forepaw movements. By feeding this motor signal to a lever controller, we could then move the metal lever, which for all purposes was now capable of reproducing, in real time, the voluntary motor intentions of the rat's brain.

A few days after the rats started to show signs that they were able to transition back and forth between either using their forepaws or their brains to control the water-carrying lever, Chapin decided to play the ultimate trick on his furry friends: he disconnected the bar from the lever. At that point, the rats would press the bar, but the lever would remain still. Frustrated, the rats would press the bar repeatedly, but to no avail. The lever would no longer move. It was then that something almost unimaginable took place.

As Chapin turned on the BMI that allowed the rat brain activity to feed directly to the lever, the rats reacted to the sudden introduction of hope into their predicament as any of us would: they tried very hard to find a way to move the lever, not by pressing the bar with their forepaws, but just by thinking about doing it.

Initially, most of these movements were cautious, and the rats were unable to reach that rewarding drop of water. Yet, a few of their attempts succeeded, and the more a rat was able to drink via this most unlikely water delivery system, the more it realized that it could reproduce a complete lever movement using its brain activity alone. Although none of the rats knew what was going on, they were producing spatiotemporal patterns of neuronal firing activity that were somewhat similar to those generated when they used their forepaws to control an intact bar-lever system. After a few minutes of interacting with the BMI apparatus, most of the rats stopped pressing the bar with their forepaws altogether. Through trial and error, the animals discovered that if they just looked at the bar and, we posited, imagined the movements needed for their forepaws to press it, they could somehow get all the water they wanted. Of course, the four rats that succeeded had no idea of the trouble we—and most especially John Chapin—had experienced to interpret and enact their thirsty motor intentions. What they appreciated was that they had become the first rats in their colony to get free drinks every time they were placed in the lab's experimental setup.

Aurora owed a great deal of her scientific success to an owl monkey named Belle, who three years earlier ushered primates into the age of brain-actuating technology.

Belle was a video game junkie, just like Aurora. Over several months, she had mastered a game that involved using her right hand to grasp a joystick while she watched a horizontal series of lights flash in front of her on a plain display panel. During her training, she had quickly figured out that if a light suddenly flashed on the screen, and she moved the joystick left or right in the direction of the light, a solenoid valve would open and a drop of fruit juice would splash into her mouth. More than any other of our owl monkeys, Belle loved to play this game. It seemed she was a juice junkie, too.

While Belle played her game, she wore a cap, glued to the top of her head with a smooth layer of surgical cement, a material normally used to repair the loss of skull bone in patients. Under the cap there were four plastic connectors, each of which received signals from rectangular arrays of Teflon-coated metal microwires implanted a couple of millimeters deep into different areas of Belle's frontal and parietal lobes. The cortical areas chosen for those implants were known to be involved in the type of visual-motor planning required for primates to translate visual cues, such as flashing lights, into the hand movements needed to move a joystick toward a target. A couple of these implants were located in Belle's M1 and other motor cortical areas. Their combined signals amounted to a small sample of the detailed motor program needed by Belle's arm and hand muscles to produce the elaborate movements that her brain had conceived in response to her video game–juice challenge. As she reflected on how to perform her motor tricks, the implants took a privileged glimpse of the sweeping electrical brainstorms that spread across her cortex as action was, spark by spark, forged out of abstract thinking.

To ensure a clear recording, each of the blunt metal tips of the microwires was nicely laid into the slightly salty, fluid-filled extracellular space that surrounded Belle's delicate cortical neurons. Strategically positioned, these naked sensors listened attentively, like a respectful and privileged confessor, to the brief electrical murmurs produced by her neurons.

Inside Belle's brain, each time one of the cortical neurons located next to an implanted microwire produced an action potential, a tiny amount of electrical current would flow through the extracellular space and be detected by the microelectrode's tip. The electrical discharges from all the implanted microwires would then feed into the microchip connected to one of the connectors situated atop the microwire arrays. This microchip, which we named the neurochip, contained the electronics needed for amplifying and filtering the tiny electrical signals produced by Belle's neurons.

From each of the neurochips, a small wiring bundle ran from Belle's head cap to an electronic closet next to the soundproof chamber in which she was playing the video game. The electronic closet, in turn, was linked to a master microcomputer that was responsible for translating Belle's thoughts into a stream of digital signals that would control the movements of two robot arms, following the voluntary motor intentions produced by Belle's brain.

But how could that translation of raw brain activity into digital motor signals be achieved? Although many in those early days thought that this hurdle would be the most difficult one to overcome, it turned out that the answer to this problem was simpler than one might have thought. I vividly recall the day that Johan Wessberg, a Swedish neurophysiologist from Göteborg who at that time was conducting his postdoctoral research in my lab at Duke, walked into my office and told me, in his calm but confident manner, that he had figured out how to implement a real-time computational algorithm for our BMI. His audacious insight, which undoubtedly opened the path to BMIs, came while he was playing around with some neuronal data, the sort that we collected day in and day out. After some careful analysis, Wessberg discovered that if he used a relatively simple algorithm, known to statisticians as multivariate linear regression and to engineers as the Wiener filter, to sum linearly the electrical activity produced by the simultaneously recorded cortical neurons, he could generate a surprisingly accurate prediction of Belle's hand position. The algorithm would be able to identify an optimal means for weighing the contribution of electrical activity from each of the recorded neurons. Then it would sum these weighted contributions to generate a continuous motor output signal that could be used to reconstruct the trajectory of Belle's wrist in a robotic device (see Fig. 6.2). It was a huge step forward from John Chapin's manual calculations.

First and foremost, Wessberg had to come up with a way to obtain a good estimate of the free arm movements he was trying to predict from monkey brain activity alone. Since we were using owl monkeys—small primates with delicate little arms—that requirement posed severe constraints on the type of technology that could be utilized to achieve precise kinematic measurements. He overcame this obstacle by adapting

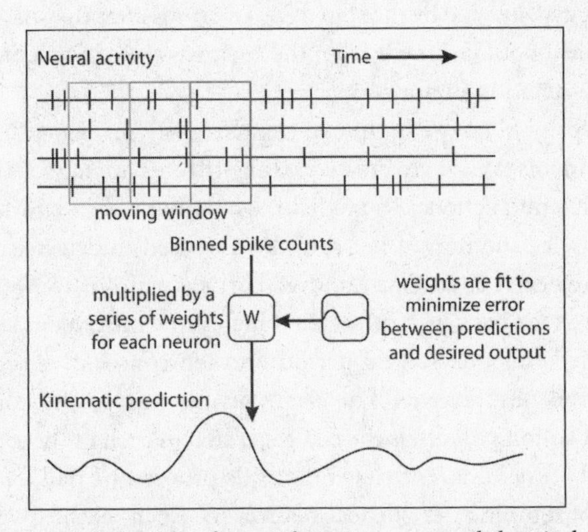

FIGURE 6.2 General algorithm for translating raw neuronal electrical activity into digital commands that can be employed to predict kinematic parameters and control artificial tools based on brain activity alone (see text for details). (*Illustrated by Dr. Nathan Fitzsimmons, Duke University.*)

a device known as the shape-tape, a flat, narrow, and very flexible plastic tape containing fiber-optic sensors, to the dimensions of an owl monkey arm. By strapping the tape along the monkey's forearm and wrist, the sensors embedded in this device could detect the monkey's arm movements; every time a monkey made a given movement, the flexible tape attached to the arm would bend somewhat proportionally. By continuously recording the bends of the tape, Wessberg could reconstruct the trajectory of Belle's wrist as she performed her motor tasks.

Having solved this first big problem, Wessberg pressed on to investigate the best computational strategy for combining the dynamic activity of a population of cortical neurons. He began by considering how the primate brain solves problems computationally to make a muscle move. His intuition was that we had to be able to handle the signals that initiate a voluntary motor command a few hundred milliseconds before any body muscle moves. Once a motor plan is generated by the brain, the large cortical pyramidal neurons transmit the commands to sizable pools of neurons in the spinal cord via their long, thick axons, which are highly myelinated and hence pretty fast conductors of electrical activity. The final product generated by the spinal cord neurons after they have received this electrical bombardment from the cortex is the equivalent of a final, ready-for-execution motor script.

Inspired by biology's highly efficient strategy for generating voluntary movements, Wessberg created an algorithmic analog for transforming thoughts into actions. To predict the continuous spatial trajectory of Belle's wrist as she moved her arm, he analyzed the electrical activity generated by each of the one hundred cortical neurons we were recording, going as far back as a full second before the movement started. He then divided this one-second period into ten consecutive segments of one hundred milliseconds. For each cortical neuron, he counted the number of action potentials the cell generated in each of these segments, also called "time bins." At the end of this process, he had a temporary database of ten bins per cortical neuron recorded, each of which contained the number of spikes produced by the neuron in a particular moment in time prior to the onset of Belle's arm movement. So, as Belle moved her arm over the course of a few minutes, Wessberg recorded her wrist's position, looking backward in time into his neuronal database every time a movement started. This component of his algorithm is known as "data binning."

The time bins were then combined into two large data sets, also known as time series: one containing a time-varying trajectory of Belle's wrist in three-dimensional space and the other containing the "binned" cortical electrical activity produced one second prior to Belle's arm reaching that particular position. These data sets were used to measure the linear correlation between the cortical neuronal firing produced by Belle's brain and her arm trajectories.

This linear correlation turned out to be highly statistically significant. After receiving the time series data, the Wiener filter returned a large number of optimal regression coefficients, each of which related to one of the ten bins describing the firing activity of a given neuron. The value of each of these coefficients directly reflected the relevance of a given bin of past neuronal firing in predicting the future position of Belle's wrist. Thus, if the regression coefficient was a high number, the neuronal firing contained in that one-hundred-millisecond bin was highly predictive of Belle's wrist position in the future. Conversely, if the coefficient was very small and close to zero, the firing activity contained in that bin had no predictive power whatsoever and could be dropped from the calculations. Moreover, if the regression coefficient was positive, the firing rate produced by a neuron at a moment in the past was directly correlated with changes in arm position in the future. If the regression coefficient was negative, the rate was inversely correlated with arm position. Using a large number of regression coefficients, we could derive a multivariate linear equation to describe the level of correlation between our sampled neuronal population firing and Belle's arm movement. We could then train the algorithm to linearly convert the time-varying activity of a sample of cortical neurons into a time-varying motor trajectory. Furthermore, after long training sessions, the coefficient values became stable. They seemed to reach an optimal level of performance. For this reason, the Wiener filter could, to some degree, artificially reproduce the kind of complex neurophysiological task normally performed by the spinal cord (see Fig. 6.3).

But Wessberg was not done. In an attempt to find the limits of his clever new computational technique, he decided to investigate whether

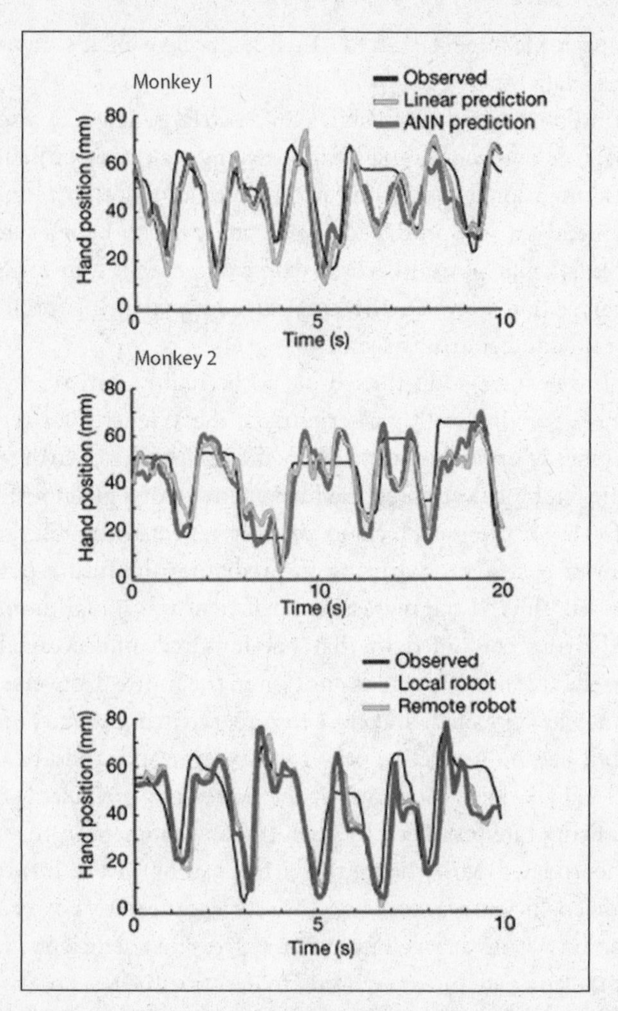

FIGURE 6.3 Where will Belle's wrist go? The graphs illustrate real-time predictions, derived from Belle's and Carmen's brain activity, and how well they reproduce the actual position of these two owl monkeys' hands. The bottom panel shows that the same kinematic predictions can be used to control simultaneously a robot arm located next to the animals at Duke University or remotely, at MIT. *(Originally published in J. Wessberg, C. R. Stambaugh, J. D. Kralik, P. D. Beck, J. K. Chapin, J. Kim, S. J. Biggs, M. A. Srinivasan, and M.A.L. Nicolelis, "Real-Time Prediction of Hand Trajectory by Ensembles of Cortical Neurons in Primates."* Nature *408 [2000]: 361–65.)*

running multiple versions of his algorithm, fed by spatiotemporal data from the same pool of neurons, could generate simultaneous predictions of multiple motor behaviors, such as the time-varying positions in three-dimensional space of the wrist, elbow, and shoulder. Against all odds, it worked. The only caveat was that, for each kinematic parameter it needed to predict, the Wiener filter had to produce a different set of regression coefficients. Still, this was a spectacular finding. Simply by mixing the patterns of cortical activity in slightly different ways using distinct weighted linear sums, Wessberg could generate multiple, simultaneous kinematic signals. There was no doubt that Belle's cortex was capable of some serious multitasking.

The next question was whether this linear computational strategy would command a robot arm to move in the skillful, biological manner of Belle's monkey arm. To find out, we had to apply Wessberg's optimal and stable regression coefficients for one thousand time bins to the algorithm. First, the algorithm would multiply each bin by its corresponding regression coefficient, which had been calculated during the training phase. Once all the necessary products had been calculated, the algorithm would sum them up, plus a fixed constant, to generate the predicted value for a particular movement in time. The operation would be repeated continuously, for each subsequent time period. If the algorithms could keep up with the data, a master computer at our lab would disseminate the motor control signals derived from Belle's brain to two computers, one located in a room next to our lab at Duke and the other at MIT in Cambridge. Each of these computers was in charge of delivering digital commands to a multijointed robot arm. As with John Chapin's rat experiments, we needed to translate Belle's neuronal activity into robot commands in just two to three hundred milliseconds—the natural temporal delay that normally separated the emergence of motor signals in the cortex and the instant in which her limb started to move.

After several months of grueling work, finally it was showtime.

As Wessberg turned on the experimental setup and lights started to flash randomly, one at a time, in front of her, Belle immediately started to move her joystick back and forth. For the next thirty minutes, which actually felt like the closest thing to the infinite that I have experienced, the master computer busily cranked out preliminary sets of linear regression coefficients until the algorithm's subroutines indicated that an optimal set had been obtained. At that point, with the coefficients changing very little, we were ready to turn on the BMI.

For a few nerve-racking, nail-biting, soul-searching moments, nothing out of the ordinary happened. Belle continued to play her game. Fruit juice continued to drop into her mouth. And the most profound silence I had heard in my life filled the room.

Without any warning, the delicate robot arm at Duke started to move its metal joints and rubber tendons. On the main computer's screen, we started to see two bright lines being plotted simultaneously. A red one tracked the gentle movements of Belle's biological arm as she played away. Then, appearing out of nowhere, a line of Duke navy blue began to plot the trajectory of the robot arm, which appeared to be trying as hard as it could to keep pace with its more senior primate counterpart. During the first few seconds, the two lines were separated by a considerable distance, suggesting that the predictions coming out of the BMI were not very accurate. Then the lines started to converge, until they almost completely overlapped.

I turned my attention to the phone, where my colleagues in Massachusetts awaited my word.

"Is it moving yet?" I said, puzzled that I wasn't greeted by an excitement to match my own.

"Nope. It is dead! Pretty dead! As dead as it could be."

"How could that be? Ours is moving just fine next door."

"Up here, nothing is happening. The arm is sitting there, frozen. Immobile, actually. Not even a jerk or a twist."

Although I was trying to keep my cool, I could see that around the room tension and uneasiness were growing steadily as everyone listened to my side of the conversation with the MIT team. Turning back to the phone, I searched for an explanation for the trouble we were experiencing. "I have no idea what is going on. It should be working. Have you checked the transmission line?"

"Yes, I have checked everything. I have gone through the checklist three times by now. It reminds me of *Apollo 13*. I should be saying, Durham, we have a problem!"

By then, I could only come up with one last idea. "Well, usually, when things like this happen at home, we start by checking whether the power switch is on."

For a second, it sounded like my interlocutor was no longer paying attention to anything I had to say. Frustrated beyond belief, he seemed instead to be talking to himself.

"I have reinitialized the computers and tried all but. . . . Now, wait a minute. Wait just a damn minute. I *have* forgotten one thing. How could I not have checked this?"

"What? What have you forgotten?" Now it was I who could not contain the angst any longer.

"I forgot to turn the robotic arm's power switch on!"

The loud and synchronous screaming coming from the phone, like the sound one hears in a packed soccer stadium when the home team unexpectedly scores a *golaço*—a really beautiful goal—told me that the arm in the MIT lab was finally moving.

At this stage in our research, Wessberg's algorithm could not come close to mimicking all of the complex physiological tasks performed by a brain's neural circuit. But it was more than enough to generate the continuous motor signal outputs to reproduce the gracious movements generated by the slender upper limbs of our owl monkey Belle.

As our MIT colleagues began to receive the motor control signals from Belle's brain, a third line appeared on our screens. Now, it was the MIT robot arm that started tracking Belle's movements in its brief but adventurous voyage into the annals of neuroscience history. Choking with the significance of the moment, I could only think of what Galileo Galilei had allegedly murmured in his own defense during his trial before the Italian Inquisition:

"Eppur si muove"—And yet it moves.

By the winter of 2002, our research team's setup was ready for Project MANE—the Mother of All Neurophysiological Experiments. It was an acronym that, not coincidentally, reminded me of my boyhood idol, Manoel Garrincha, the star of the 1958 and 1962 Brazilian soccer teams that, in winning back-to-back World Cups, set new standards of excellence for the beautiful and most popular sport in the world. Born with severe bone deformations that made his knees and legs point and bend in different directions, Garrincha learned to take advantage of this unusual conformation to create a dribbling technique—full of hip twists and body fakes. He taught his teammates, by example, to dance their way through the game, creating an exquisite ballet in which the ball became an extension of the player's feet. Garrincha's nickname, the name by which he is known to all Brazilians, is Mané.

The MANE experiment was far from trivial to design, implement, and execute. Several computational tasks and engineering components, each of which contained many new procedures and parts, had to work flawlessly, in real time, in order for the data we collected to be of use. Besides, like anything that has never been tried before, it was difficult to predict the outcomes of the experiment. Moreover, if not for the strong and visionary support of Alan Rudolph, an accomplished scientist who worked as a program director at the U.S. Defense Advanced Research Projects Agency (DARPA) at the time, we would have never have mustered the funding needed to run MANE.

After months of assembling large pieces of hardware, generating new computer programs, fixing computer glitches, and convincing Aurora to play video games, of all things, the time for the first complete test had arrived.

As we prepared to "fire up" the experiment, our conversation was sporadically interrupted by Aurora's voice, pestering us to hurry up. Nobody could blame her for being a tiny bit impatient. After weeks of intense training, several promising attempts to run the experiment had been recently aborted. For a change, that night our computers, spread out on a large lab bench in the experiment's control room, did not crash and eagerly swallowed the gargantuan amounts of data they were being fed.

Aurora was, as usual, ensconced in her favorite chair, equipped with a fancy joystick, a flat-panel LCD screen, and a fruit juice dispenser, so that she could drink as much juice as she wanted as she went through the experiment. Nearby, a 512-channel multineuronal acquisition processor (MNAP), the largest machine of its class in the world, sat equally ready to record Aurora's brain activity. For this run, we utilized only ninety-six of its channels. The other members of the team—Jose Carmena, a hyperactive former electrical engineer and Spaniard who loved to operate multiple microcomputers, each fitted with two monitors, and Mikhail (Misha) Lebedev, a Russian physicist turned neuroscientist with a soothing capacity for problem solving—were poised around the bunker. Both were big soccer fans—that was a given. But Lebedev also brought a particularly useful skill to the experiment: he had worked with Aurora in her previous job and knew firsthand how difficult she could be.

Things were moving steadily in the control room. While Carmena frantically circulated among the computers, checking their status, I continuously observed Aurora's "body language" through a series of TV monitors, each of which displayed the view from a different video camera.

Behind the control room, we had placed a sturdy industrial robot arm. The arm itself had a freedom of movement of seven degrees; the rudimentary hand at its end contained just two fingerlike appendages capable of grasping simple objects and holding them secure. The robot lay immobile, its artificial shoulder and elbow partially stretched, its hand fully opened—a beautiful piece of machinery, almost pleading for someone to take control of its inert joints and motors and generate purposeful and coordinated movements, capable of accomplishing something *meaningful*.

Cooing gaily, Aurora—or, I should say, Aurora's brain—prepared to unleash the unimaginable. In previous attempts Aurora had not shied from letting us know that if the robot arm was not working, the fault was not hers. Tonight, however, she was performing without complaint. Lebedev entered the lab and, sliding past the cables and hardware next to Aurora's chair, whispered something in Russian into her ear. He never told us what he said, but we became convinced that it inspired her. Staring intently at the computer screen in front of her, Aurora made faces at one of the cameras recording her. She was having fun, we could see. Yet, her inquisitive eyes conveyed that she was running out of patience and wanted to start the experiment.

As soon as Lebedev closed the door of Aurora's chamber, Carmena gave the "go" to start MANE. At that moment, the cumulative years of research and the hopes of thousands of severely paralyzed people who dreamed of one day regaining some degree of their former mobility became deeply intertwined.

We knew very well that most of the experiment depended on the work of the MNAP, a few computers, and—mainly—Aurora's very own stubborn brain. As she gently grabbed the joystick with her left hand and applied her intriguing and charming voluntary motor intentions to her favorite video game, our attention was glued to the sleek, swift blinking matrix on the computer screen that represented the flow of electrical activity produced by the population of ninety-six cortical neurons we were recording. I had also rigged up a loudspeaker in the control room, so that I could listen to her brain symphony, which sounded something like a violent thunderstorm of sheer beauty as it crisscrosses a soft tropical summer night sky. Lost in a contemplative awe akin to the amazement one unexpectedly encounters in the presence of a miracle created by nature, we were gripped by the intimacy and revelation that Aurora had so generously granted to us.

The sample of Aurora's motor thinking was obtained simultaneously from multiple cortical areas of her brain. In the previous three decades, studies carried out by many laboratories had identified which of the cortical areas in the frontal and parietal lobes are involved in the generation of the type of neural motor plans required for someone like Aurora to produce precise arm and hand movements. According to the rules of the experiment, Aurora needed to use such precise arm-hand movements to manipulate her joystick and control the two-dimensional trajectories of the cursor on her computer screen. By moving this cursor, Aurora could intercept and grab the target object—a large, filled circle—which randomly appeared in one of a variety of locations on the screen at the beginning of each experimental trial. Using her left arm and hand, Aurora had learned to play the video game quite well, particularly because, if she succeeded and grabbed the target in less than five seconds, she received a very satisfying reward: a drop of her beloved fruit juice. Thus, every time Aurora got the game right, a high-frequency sound, produced simultaneously with the opening of the solenoid valve that delivered the juice to her mouth, filled the control room (see Fig. 6.4).

As Aurora started to play the game, we were able to use the MNAP to record the spatiotemporal patterns of electrical signals produced by the

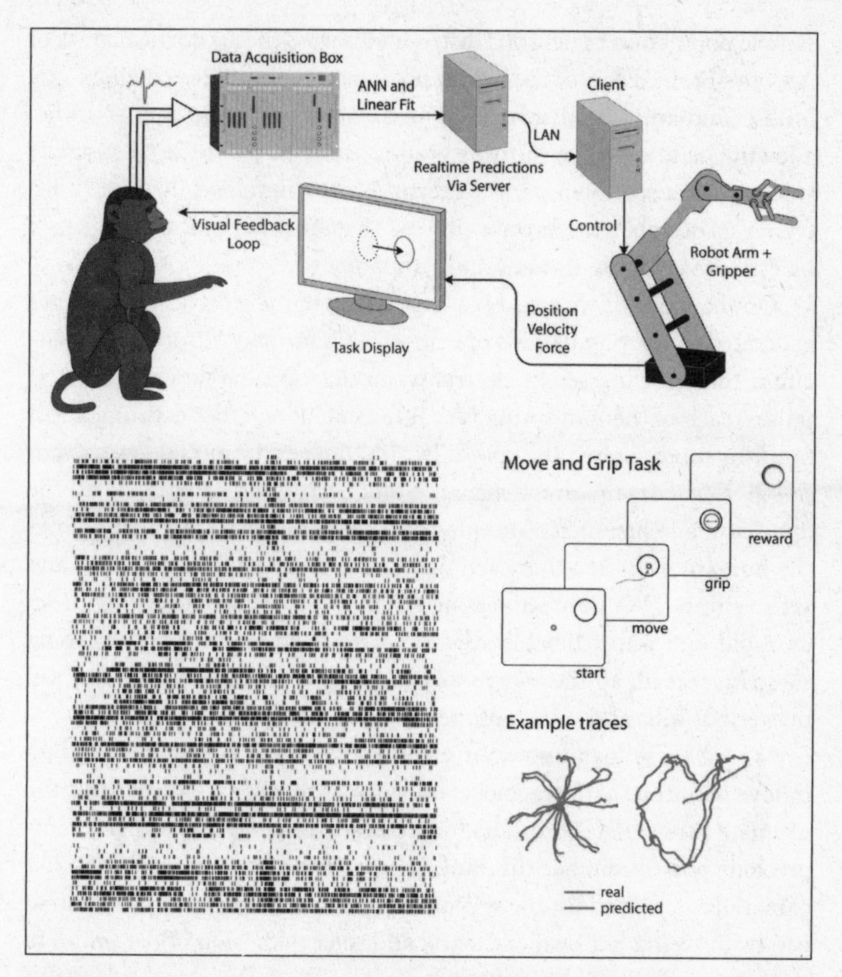

FIGURE 6.4 Aurora loses her joystick and frees her mind. On top, the experimental setup employed for Aurora to use her brain activity alone to control the movements of a robotic arm. On the bottom left, a sample of the electrical activity of 96 cortical neurons from Aurora's brain. On the *Y* axis, each vertical bar represents an action potential produced by a single cortical neuron. The *X* axis represents time (10 seconds). In the right panel, a graphic representation of the task Aurora performed and examples of predictions of arm movements based on combined brain activity. *(Originally published in J. M. Carmena, M. A. Lebedev, R. E. Crist, J. E. O'Doherty, D. M. Santucci, D. R. Dimitrov, P. G. Patil, C. S. Henriquez, and M.A.L. Nicolelis, "Learning to Control a Brain-Machine Interface for Reaching and Grasping by Primates."* Public Library of Science *1 [2003]: 193–208.)*

sample population of neurons distributed across the six cortical areas of Aurora's brain. Each of the action potentials typically lasted about one millisecond and was then plotted on a computer in the control room, allowing us to monitor Aurora's brain activity in real time. We eagerly tracked these spatiotemporal patterns as they unfolded in front of us, trying to decipher the precise process that Aurora's brain used, like a maestro, to compose its neuronal symphony.

On the monitor we could see large waves of the electrical discharges generated by the populations of neurons that formed Aurora's brain circuits. That brewing sea of electricity, an unstoppable flow of electrical pulses, each of them meaning very little, but all together defining a vast brainstorm, registered the epic of her lifetime: every motion, every sensation, every dream, every memory, and every sorrow, but also all the joy that made her such a unique character. The very stuff that defines the human brain, making each member of our species so unique and yet so similar. We were witnessing the fluid life cycle of thinking, from its rapid and unpredictable seed at birth, to its waving and warping rampant spread, all the way to its agonizing disappearance, in the still mysterious hills and valleys of the cortical mantle.

We sat there for minutes, irremediably and unconditionally falling in love with the small fragments of Aurora's thoughts. Now we had the ultimate proof that Aurora had joined our team. By sharing her most precious possession, her thinking, Aurora was not only displaying her camaraderie toward us; she was offering an unrivaled example of altruism by devoting her own existence, and the exploration of her brain, to the future benefit of millions of people.

But we needed to get to work to ensure that solemn offering was not made in vain. Our first goal was to sample and record the activity of as many of Aurora's cortical neurons as possible while they were contributing to the symphony of motor control signals involved in playing her video game and winning liters of fruit juice. Then, we had to run the electrical signals generated by Aurora's brain through a series of simple mathematical models and extract from these signals the motor commands required for Aurora to move her left arm and hand. These included, for instance, the continuously changing spatial position and velocity of her wrist, elbow, and shoulder, as well as the gripping force she applied to the joystick handle. The commands could also be derived exclusively from linear combinations of her neuronal activity—a very simple calculation, given the parameters involved.

Thirty minutes into the experiment, the first good news came: we realized we could produce very accurate real-time predictions of the cursor movements that Aurora needed to generate in order to reach and grab a new circular target as it appeared on the computer screen—just with our ninety-six-neuron sample of Aurora's brain activity. In other words, we could reproduce the type of movements Aurora used to get her juicy reward.

But this was nothing compared to what came next.

While the sounds rendered by the electrical discharges of Aurora's brain filled the control room, we slowly recognized that changes were occurring in the firing frequency of her neurons even before she started to move her arm or hand. We could recognize which movement Aurora was planning to make long before she actually contracted any muscle in her arm—a few hundred milliseconds before any movement happened. The brain signals that we were hearing contained enough of the motor plan needed for Aurora to play her video game.

As Aurora's movements became more and more precise, her performance improved, to the point that she was reaching and grabbing the circular target virtually every time she played a new trial of the video game. And as Aurora improved, so did our mathematical models. By now, they were converging into an optimal performance, meaning that, using only brain-derived signals as inputs, our models could generate time-varying outputs that predicted, amazingly well, what arm and hand trajectories Aurora's brain was going to execute, before Aurora moved a single muscle in her arm.

Next, we started routing the outputs of these models to the robot arm located in the control room. After a few seconds of indecision—apparently even the robot arm could sense the anticipation of the moment—the machine began to generate movements that mimicked those being produced by Aurora's left arm and hand. Aurora's motor thoughts were now directly controlling not only her own arm but also the robotic one, which imitated its biological counterpart. And just as easily as her arm followed the commands of her brain, the robot arm followed them, too. Indeed, there was not much disparity between her voluntary motor intentions and the robot arm's performance. If anything, the robot was a bit faster in materializing its mistress's desires.

Progressively, the movements of the robot arm became more precise, which presented an opportunity to see exactly how powerful Aurora's motor commands were. Without a second thought, Lebedev went to the experimental room and gently removed the joystick from Aurora's reach. After wishing her good luck, he left her alone, with the game still playing on the computer screen. Next, Carmena switched the control over the cursor movements, from Aurora's old joystick to the robot arm's wrist, which was commanded by the BMI. That meant that from now on, to move the computer cursor, intercept and grab the target, win the video game, and get her rewarding drop of fruit juice, Aurora had only one option. Neither the joystick nor her own arm movements would do the trick any longer. Instead, she would have to perform this complex motor task by operating a BMI.

As with Belle's BMI setup, this BMI had been created to generate upper-limb motor movements just by thinking. Aurora would have to *imagine* the arm movements she wanted the robot to use to intercept the target. Her thoughts would guide the robot arm, rather than her own limb, to direct the cursor to the exact spot on the screen where she could grab the circle. Of course, that circle, though it was right in front of her, was also a virtual object, so Aurora would also need to imagine the arm touching a ball in space.

We had not trained Aurora to handle this part of the experiment; she had to figure out how to operate the BMI on her own. Things had changed dramatically. At first she looked puzzled and surprised. After hesitating, and then committing a few errors, Aurora rose to the occasion better than any previously rejected and discredited scientist. Cooing to tell us that she had understood the idea, she relaxed her long, muscular arms, resting them around her chair. After taking aim at the screen, her attentive eyes fixed on the cursor, waiting for a new circular target to pop up. Back in the control room, we began to hear and see what many scientists had told us would be impossible: as the initial salvos of a brainstorm charged across her cortex, the sound and flashes produced by hundreds of action potentials filled the room as Aurora's motor intentions were decoded in real time by our mathematical models. Even before Aurora could visualize the end result of her thoughts, our BMI was transferring the motor commands extracted from her brain activity to the robot arm. Just as a new target appeared on the screen, the robot arm started to move, seeking in the empty space of the control room an elusive and unseen object whose location had registered only within Aurora's eyes and brain. Back on the screen placed in front of Aurora, the computer cursor—now under the control of the robot wrist—slid into a beautiful and intentionally curvy trajectory aimed right at the target's center. That first target, and the many targets that followed, was grasped with almost human gusto by the mechanical arm, whose graceful movements had been generated by voluntary brain activity alone. At long last, Aurora's brain had been set free from the limiting constraints of her biological body.

Aurora was now playing her video game just by thinking. No need to use her own arms anymore. Her brain activity, free of her body and self-sufficient, was carrying out, across laboratory walls, every bit of the burden generated by her voluntary will. But something even more stunning was unfolding right in front of us. As Aurora's neurons directly controlled the movements of the robot arm, her brain began to assimilate that piece of machinery into the neuronal image of her body, almost as if the robot had become an extension of her self.

Aurora's achievement with the BMI demonstrated a third principle of neural ensemble physiology, the *distributed coding principle.*

THE DISTRIBUTED CODING PRINCIPLE

Any type of information processed by the brain involves the recruitment of widely distributed populations of neurons.

Since Aurora's achievement, this principle has been validated in an extensive series of experiments, carried out by many laboratories worldwide, that have measured how brain circuits process information related to perceptual, motor, and cognitive tasks. Essentially, by sampling populations of neurons, distributed across vast areas of Aurora's brain, we were able to mathematically translate the probabilistic nature of her neurons into deterministic motor behavior.

As we saw in chapter 5, the flavor of distributed coding that I envision involves a dynamic understanding of spatiotemporal receptive fields and cortical maps. As such, the regional functional specialization of cortical areas, initially determined by the way the cortex was built during early development, is probabilistic. That also means they never commit 100 percent of their allegiance to any particular function. Thus, although neural circuits spatially distributed in the S1 cortex may have a very high probability of firing in relation to a particular tactile stimulus, the probability that they may also fire in relation to a stimulus from a different sensory modality is not zero. Moreover, in certain contexts, particularly those involving changes in body constraints—such as Aurora's loss of control over the video game by using a joystick—neurons may fire robustly in response to stimuli other than those to which they were originally assigned. Any change in the original body constraints (e.g., blindness), modifications in experience (e.g., learning to play the piano), or increase in task demands (e.g., playing in the Majors instead of in triple-A ball) will readily change the shape of these distributions and reallocate functions across the cortex. I propose that the notion of distributed processing is a *universal coding strategy* employed by the entire neocortex, being valid either for local subregions or across its whole.

Because of this neurophysiological distributed coding, we believed far-reaching opportunities would immediately present themselves once we were able to demonstrate unequivocally that a BMI could generate movements in a robotic device controlled by brain activity. We proposed that BMIs would lead to the development of a new generation of neuroprosthetic devices aimed at restoring the mobility of millions of patients suffering from devastating levels of body paralysis, including that caused by the ALS that had conquered Dr. César Timo-Iaria's body. In addition to providing a completely new experimental tool to probe the basic mechanisms that govern the operation of brain circuits, the Mother of All Neurophysiological Experiments had fulfilled its goal of making this a credible clinical possibility to be studied.

Back in the control room, the high-frequency noise of the solenoid continued to burst unimpeded as Aurora got drunk on an ocean of fruit juice. By the bright look of her face, it was clear she was savoring each millisecond of her triumph over the unexpected task we had presented to her. Indeed, when we had finished cheering and hugging and had calmly returned to our chairs in the control room, I could almost swear that Aurora flashed those shiny and mischievous black eyes away from her computer screen to stare at us, through one of the video cameras, and, flirtatiously, as only she could, gave us a wink.

Apparently, no one but me saw that wink. Both Lebedev and Carmena promptly argued that such a thing was utterly impossible. But then, everything we had accomplished that night had once been deemed impossible. Therefore, I would not be surprised, if on detailed analysis, our videos one day reveal that Aurora did in fact manage to pull off yet another of her rhesus monkey tricks.

7

SELF-CONTROL

In the mid-1960s, a handful of scientific reports surfaced indicating that human subjects were capable of achieving exquisite voluntary control of the muscle fibers innervated by the axon of a single alpha-motor neuron located in the ventral horn of the spinal cord. This launched the era of biofeedback research. The people who had participated in these studies had exhibited this level of control over Sherrington's single "motor unit" by using visual or auditory feedback of the activity recorded by electrodes inserted into their muscles, conveyed using either a light pulse on a monitor or a sound pulse from a speaker. After just fifteen to thirty minutes of training in these biofeedback experiments, most human subjects became highly proficient. Further training allowed the same people to repress the motor unit they had earlier recruited and select another one for voluntary control.

Around that same time, Drs. James and Marianne Olds, then at the University of Michigan, found that they could artificially increase the firing rate of individual neurons, located in many distinct sensory and motor cortical areas in animals, by stimulating the brain's reward-pleasure system in lightly anesthetized rats. In their experiments, every time a single neuron they were recording produced an action potential, the rat received an exogenous electrical stimulus directly to a brain structure that generates powerful hedonic sensations, such as those associated with eating or mating. Accordingly, every time the monitored cortical neuron fired, the rat received some near-orgasmic biofeedback. The Oldses observed that this ingenious reinforcement loop resulted in a significant increase in the neuron's firing rate.

Inspired by these studies, the German-American neurophysiologist Eberhard Fetz decided to apply biofeedback to his own innovative experiments with primates. After working under the renowned pain neurophysiologist Patrick Wall, and graduating from MIT's physics department, Fetz had joined the Department of Physiology and Biophysics and the Regional Primate Research Center at the University of Washington in Seattle. A promising assistant professor, he also learned the new art of recording single neurons in behaving monkeys. Moreover, by collaborating extensively with research psychologist Dom V. Finocchio, an expert in operant conditioning, Fetz gradually realized that there was yet another way to investigate the physiological properties of cortical neurons.

In one of his first studies at the University of Washington, published in *Science* in 1969, Fetz used his new technique to investigate individual neurons in the primary motor cortex of awake rhesus monkeys. Although a few scientists were dismissive of Fetz's approach, the experiment laid the early foundation for the creation of brain-machine interfaces three decades later.

Like all leading primate neurophysiologists of his time, Fetz started by monitoring, for a few hours each day, the extracellular electrical activity produced by a single neuron using a solitary tungsten microelectrode. Once that neuron was recorded, the microelectrode was slowly lowered a few hundred microns into the M1 cortex using a hydraulic microdrive. But such serial recording was the only orthodox aspect of Fetz's experimental approach. In a strike of audacity and ingenuity, he decided to link the amount of food reward his monkeys received to the animal's ability to generate high levels of firing from the single cortical neuron he was recording. This meant the level of the animal's own brain activity dictated how much reward it could gain. The reward may not have been as orgasmic as a microstimulation of the pleasure center, but it was still pretty good: a banana-flavored food pellet.

Fetz's experimental apparatus combined both operant conditioning and biofeedback. After isolating a single neuron in the M1 cortex of a monkey, Fetz recorded the action potential produced, and each time the action potential crossed a particular voltage threshold, a trigger mechanism produced a voltage pulse. These voltage pulses were combined by a simple resistor voltage integrator, which Fetz called the "electronic activity integrator." When the integrator's voltage hit a high enough total level, a feeder released the food pellet near the monkey's mouth. This created a direct link between neuronal firing rate and food reward.

To assist his subjects in gaining their banana-like feast, Fetz provided auditory and visual feedback that indirectly signaled the level at which their monitored cortical neuron was firing at a given moment. After a few training sessions, all of the animals learned to associate high-intensity auditory clicks or illuminated meter movements with the imminent delivery of a delicious meal. Shockingly, he had discovered that nonhuman primates could, like humans, learn to voluntarily control the firing of individual M1 neurons.

On close inspection, Fetz saw that the monkeys were able to spur a burst of firing in the monitored neuron for one hundred to eight hundred milliseconds, which, as he wrote in *Science*, "were sometimes accompanied by specific, coordinated movements such as flexion of the elbow or rotation of the wrist." But only "sometimes"! Fetz emphasized in later papers that often he would record a neuron in the motor cortex for which an increase in firing rate did not generate *any* discernible muscle contractions. More puzzling, this neuron was invariably surrounded by other cells that usually did fire when a specific muscle was contracted.

To uncover what was happening in those cases, in the early 1970s Fetz and Dom Finocchio decided to enhance the electronic activity integrator. In this upgraded version, the typical primate chair was rigged so that a monkey's head was restrained from moving and its left arm was embedded, in a semi-prone position, in a cast molded to the shape of the arm. So that the monkey would still be able to produce isometric arm-muscle contractions (generating muscle force without changing the muscle length or the angle of the joint associated with it), the cast was locked with the animal's elbow fixed at 90 degrees and its wrist and fingers fully extended at 180 degrees. Pairs of braided stainless-steel electrodes, inserted in each of four arm muscles, allowed for continuous electromyograph (EMG) recordings.

The most substantial change introduced, however, was the way in which the electronic integrator itself operated. Now, instead of feeding the integrator with only the electrical activity of a single neuron, Fetz and Finocchio added a few new inputs to their device: voltage pulses corresponding to the electrical activity generated by each of the four individual muscles being recorded. In this arrangement, the contributions of each of the inputs—whether a single muscle or the single neuron—could be weighted. This meant that Fetz and Finocchio could change the relevance of the activity of a given muscle (or neuron) in determining the final voltage level of the integrator—and what electrical activity would garner a food reward for a monkey.

After determining that their monkeys could handle this modified apparatus, Fetz and Finocchio investigated what happened when a single muscle, a single neuron, a subgroup of muscles, or even different combinations of muscle and neuronal activity were used as the main input source to the integrator. From the start, the monkeys, which continued to receive auditory and visual feedback cues, typically tried to reach the desired high voltage by concurrently contracting all muscles at once. They understood the new rules of the game and were trying to outsmart their primate cousins. To direct their subjects' behaviors, Fetz and Finocchio changed the feedback mechanism so that a set of colored lights was used to indicate which of the individual muscles had been contracted. This allowed the researchers to reinforce the contraction of a particular muscle while eliminating the other muscles from the integrator's voltage sum. The monkeys soon recognized they would only get their food pellet if they contracted the chosen muscle.

As each monkey learned to limit its voluntary motor intentions to a single muscle, Fetz and Finocchio continuously recorded the activity of one of its single M1 neurons. They found that most of the single neurons in the monkey M1 cortex fired in response to passive movements of the joints of the animal's contralateral arm in the absence of EMG activity, which meant that sensory information coming from the body periphery affected most neurons in the motor cortex.

When Fetz and Finocchio started to use the apparatus to reinforce isometric contractions in each of the four distinct muscles, they noticed that the cortical neurons they recorded exhibited some unexpected properties. For example, a large number of the single neurons seemed capable of modulating their firing rate prior to or during the contraction of multiple muscles. Indeed, some neurons were coactivated, either with the same or with different firing intensities, with all four muscles. As with the dynamic receptive fields of rat whiskers, Fetz and Finocchio had demonstrated that individual M1 neurons fired before the contraction of several muscles and that the correlations between neuronal firing and muscle contraction changed dramatically, depending on the context and type of movements being generated. As a result of these findings, Fetz later coined the term *muscle field* to define the set of muscles coactivated by the firing of a single cortical neuron. This muscle field undermined the idea that parallel "labeled lines" linked individual M1 neurons to specific individual body muscles.

The most astonishing result to come out of Fetz and Finocchio's experimental setup was obtained when they used the weighting options provided by the electronic activity integrator in an attempt to dissociate the observed correlations in the firing of individual neurons and individual muscles. To do this, they simply reinforced high firing rates of a single cortical neuron while suppressing the muscle activity in the final integrator voltage. After only a few minutes the monkeys were capable of selectively raising the firing rate of only the neuron, without generating any simultaneous electrical activity (and contraction) in the muscle field, including the single muscle most strongly correlated to that particular M1 neuron. Through selective reinforcement, Fetz and Finocchio had trained their monkeys to completely dissociate cortical neuron activity from peripheral muscle contractions.

Not yet satisfied by their evidence, they next tried to obtain the opposite effect, by selectively reinforcing muscle contraction while suppressing the firing of a single neuron. Even though the neuronal-muscle pair chosen for this experiment (an M1 neuron and the biceps) exhibited a very strong correlation in firing, and despite the fact that at the time of testing their subject was tired and totally satiated with banana-like pellets, Fetz and Finocchio reported a 10 percent suppression in neuronal firing accompanied by a 300 percent increase in bicep activity. Despite the fact that they were monitoring only one cell in the primary motor cortex at a time, and so could not document the full dynamic activity of M1, they had powerfully shown that it truly was possible to dissociate the mind from the body's flesh.

With this series of experiments, Fetz and Finocchio unveiled a much more malleable relationship between M1 neuronal firing and muscle activity. Indeed, in one of those peculiar coincidences that sometimes happen in science, almost exactly a hundred years after the original discovery of the motor cortex by the German scientists Eduard Hitzig and Gustav Fritsch—a finding that became a decisive plank for the localizationists—a team led by a young German-American physiologist now claimed that bursts of action potentials, produced by single neurons located in the same primary motor cortex, did not necessarily lead to muscle contractions at the body periphery or dynamically change the control exerted over multiple muscles. Cortical function was not so localized and predetermined after all. There was plenty of room for flexibility and fine adjustments, even in the motor cortex.

While Eb Fetz had been perfecting the pleasurable delivery of banana-like food pellets just by thinking, a few laboratories had been experimenting with the possibility of conditioning animals and people to be able to enhance rhythmic neuronal activity in the brain. Using a variety of operant conditioning techniques similar to those utilized by Fetz and Finocchio in their monkeys, as well as the noninvasive scalp EEG, several neurophysiologists recorded rhythms in the visual and sensorimotor cortices to provide subjects with biofeedback of their ongoing brain activity. Pleasant sounds, flashing lights, and even the projection of pleasant images were commonly used to reward humans when they produced the EEG rhythmic activity requested.

Using variations on this basic approach, researchers including Joe Kamiya at the Langley Porter Neuropsychiatric Institute in San Francisco reported that human subjects could not only learn to control a specific motor unit, they could learn to control the appearance of their alpha rhythm, an eight-to-thirteen-hertz oscillating activity that usually appears in the visual cortex when a subject closes the eyes and settles into a state of relaxation. Similarly, M. Barry Sterman and his collaborators in the Department of Anatomy and Neurology at the University of

California, Los Angeles, found that cats learned to control the production of their mu rhythm, a seven-to-fourteen-hertz oscillating activity detected over the sensorimotor cortex when the animal is idle or has ceased all limb movements, when they were reinforced to do so with a reward of either food or a pleasurable direct brain stimulation.

Things got so wild that one author, Edmond Dewan, said that he had become so good at controlling his alpha rhythm activity that he could send Morse code messages to a computer using his own EEG.

Although at the time neither Fetz nor the researchers using EEGs could have predicted it, this innovative work with biofeedback split the scientific community in a very peculiar way. Unlike the intellectual disputes over whether a single neuron or a population of neurons served as the basic functional unit of thinking, this divide involved those who, like Fetz, utilized "invasive" intracranial recordings to obtain brain signals and those who preferred to rely on "noninvasive" methods, such as the scalp EEG. The fundamental differences inherent in these two general approaches for sampling brain activity define the almost insurmountable chasm between those who pursue invasive and noninvasive brain-machine interfaces today.

The first person to formulate how biofeedback might one day blossom into the invention of BMIs was not one of the pioneers of biofeedback but yet another cortical neurophysiologist employed by the National Institutes of Health. In a paper published in the *Annals of Biomedical Engineering* in 1980, Edward Schmidt presented a scientific manifesto for the creation of a new field aimed at, among other things, building a new generation of prosthetic devices designed to restore mobility to severely handicapped patients.

Unfortunately, Schmidt's daring proposition immediately hit two major walls: one theoretical and the other experimental. Just a few years later, Apostolos Georgopoulos and his collaborators categorically demonstrated that the firing of a single neuron in the motor cortex was not accurate enough to generate an unambiguous prediction of the direction in which a monkey intended to move its arm. A population of neurons was needed in order to produce an accurate prediction of a time-varying primate arm trajectory in space. To transform Schmidt's vision into clinical reality, neurophysiologists would need to record the extracellular activity of large numbers of single neurons simultaneously—and there were few, if any, neurophysiologists in the mid-1980s who believed that such neural ensemble recordings could be obtained in the foreseeable future.

By the time John Chapin's rats in Philadelphia and our monkeys at Duke started to operate the sort of gizmo that Schmidt had envisioned, almost twenty years had elapsed.

Starting in 1998, a series of scientific breakthroughs helped pave the way to Aurora's mastery over the robot arms. That year, the neuroscientist Philip Kennedy and the neurosurgeon Roy Bakay, both then at Emory University, reported the case of a patient suffering from a variation of the "lock-in syndrome," a neurological condition in which subjects are completely paralyzed but still have an intact or partially functional central nervous system, who could move a computer cursor with the activity of a single cortical neuron. After some consultation, the patient had consented to an experimental device, called a cone electrode, to be implanted in his cortex. The electrode was designed to record the activity of neuronal processes that would, theoretically, interweave with the electrode's surface. Not much raw data was disclosed in that original report, and subsequent experiments in animals and other patients did not support the cone electrode's efficacy. Yet, it was clear that the time was fast approaching for the translation of basic neurophysiological techniques and concepts into clinical applications.

Indeed, a year later enthusiasm grew when Niels Birbaumer of the Psychology Institute at the University of Tübingen, Germany, reported in *Nature* that locked-in patients had been taught to control a computer-aided spelling system using their EEG rhythmic activity. Using this early "brain-computer interface" system, patients became able to write letters and e-mails—in some cases, communicating with their loved ones and the external world for the first time in years.

Shortly afterward, John Chapin, his students, and I published our study showing that rats could learn to use their brain activity to directly control a robotic lever. Although our rats rapidly learned to slake their thirst, it was obvious to us that the future of BMI research lay in primate experiments.

When we first began to plan our experiment with Belle and her owl monkey friends, Johan Wessberg and I had conservatively thought that getting the monkey to move a joystick to the left or to the right in response to a visual cue would be the most we could expect to achieve. During those early wobbly days of BMI research, we, along with most of the other handful of neuroscience groups trying to make their way into this emerging new field, would have been thrilled beyond belief if someone could have assured us that some species of monkeys, let alone our cute little owl monkeys, would get this far in their first shot at controlling an upper-limb BMI apparatus. So Belle's stellar performance in her first contact with our BMI immediately raised the feeling in all of us that she could handle something trickier.

This impression led Wessberg to test whether our newly unveiled BMI-controlled robot arm could reproduce the type of free arm movements that monkeys use in the "real" world. At first glance, this sounded like quite a challenge for a gentle owl monkey, born somewhere in the idyllic canopy of the South American rain forest and now sitting in a laboratory bunker in North Carolina. Yet, following our instincts, Wessberg and I believed we could generate much more natural, and consequently much more variable, arm movements in our experiments.

Looking at how eagerly and inquisitively Belle went for the joystick every day when she was brought to the lab setup with its juice dispenser, Wessberg sensibly figured out that the real-world task that we should study was food grabbing. In this new task, our monkey would first have to sit attentively in a chair and, for a few seconds, face an opaque Plexiglas barrier. Suddenly, the barrier would lift, revealing a square tray with a juicy chunk of fruit—Belle's favorite meal treat—placed enticingly in one of its corners. Upon seeing the fruit, Belle would have to use her right arm to reach, grab, and bring back the food to her already salivating mouth, and do so before the barrier came mercilessly down. If all this happened as planned, the monkey would have completed a single task trial correctly and been treated to a fair reward. Then, as soon as that delicious fruit chunk had been gobbled up, Belle would be ready to start her search for gustatory pleasure all over again, until the moment when, fully satiated, she would do what every wise primate usually does after such a feast: she would take a nice afternoon siesta.

As we had done in our earlier experiment, we wanted to demonstrate that our BMI could translate the raw electrical activity produced by about one hundred cortical neurons from Belle's brain into meaningful, three-dimensional robot-arm movements that closely matched those produced by the monkey reaching for the fruit. This is exactly what happened. As Belle used her arms to reach toward the fruit chunk, so did the robot arms at Duke and MIT. As the multijointed robot arms accurately mimicked Belle's arm based on a ridiculously small, random sample of neurons, it was difficult not to ponder the implausibility of what we were witnessing: the robot arm's trajectories were about 70 percent similar to the monkey's free arm movements. Based on these results, 95 percent accuracy could be achieved with a much smaller population of neurons than we had ever imagined. The question remained, how many?

The neurophysiological data obtained in further experiments revealed another important finding. Since in Belle and another owl monkey, Carmen, multiple cortical areas had been implanted with a microelectrode array, we were able to measure the contribution of individual neurons as well as cortical ensembles to the real-time prediction of arm movements in the BMI. To quantify this relationship, Wessberg created yet another new analytical technique, now known as the neuron-dropping curve (NDC) (see Fig. 7.1), which measures the accuracy of a particular BMI's computational algorithm in predicting a given motor parameter as a function of the number of single neurons recorded simultaneously. NDCs are computed by first measuring the performance of the entire sample of simultaneously recorded neurons from a given brain region. Once this "maximum" performance is obtained, the same calculations are repeated after individual neurons are randomly dropped from the original sample. This step of randomly dropping neurons is repeated many times until the overall population is reduced to a final, single neuron. Figure 7.2 shows pairs of NDCs highlighting the contribution made by populations of neurons, located in two different cortical areas (primary motor cortex and posterior parietal cortex), to the simultaneous prediction of two time-varying motor parameters during operation of a

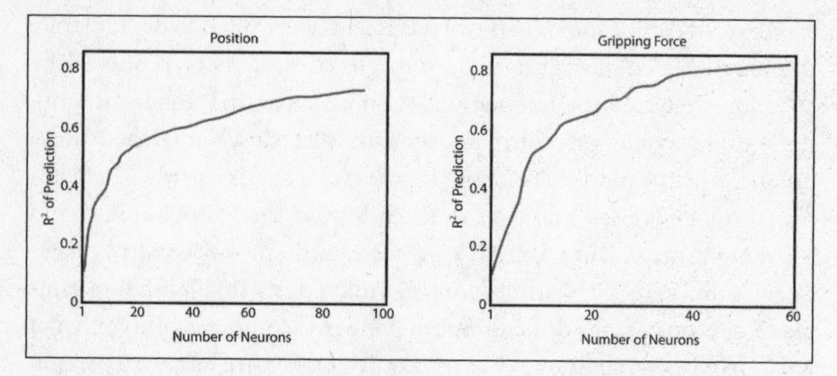

FIGURE 7.1 How many neurons does it take? Two neuron-dropping curves correlate the number of neurons (*X* axis) with the accuracy of real-time prediction obtained for two distinct parameters (hand position and hand gripping force). The same sample of neurons from the primary motor cortex of a monkey was used to construct these curves. *(Originally published in J. M. Carmena, M. A. Lebedev, R. E. Crist, J. E. O'Doherty, D. M. Santucci, D. R. Dimitrov, P. G. Patil, C. S. Henriquez, and M.A.L. Nicolelis, "Learning to Control a Brain-Machine Interface for Reaching and Grasping by Primates."* Public Library of Science *1 [2003]: 193–208.)*

BMI by a rhesus monkey. This figure illustrates how the predictions of two such parameters, hand position and gripping force, varied as a function of the size of the recorded neuronal population.

Simple as they are, the development of NDCs allowed us to obtain many useful comparisons from our recorded neurons. For starters, we were able to compare how well neural ensembles located in different cortical areas predict a particular motor parameter or overall behavior. We also reaped a quantitative measurement of the performance of neural ensembles of different sizes as well as of the effect produced by combining all simultaneously recorded neurons, irrespective of their anatomical location. Furthermore, we now had the means for detecting the average contribution of individual neurons, and for comparing the performance of similarly sized populations of neurons, located in distinct cortical areas, in predicting a variety of animals' behaviors, such as one-dimensional versus three-dimensional arm movements. But since nothing is perfect in this universe of ours, NDCs also posed a major inconvenience for a scientist working on a tight research budget: they required significant computational power (or a lot of patience) to be calculated. At the time, not having

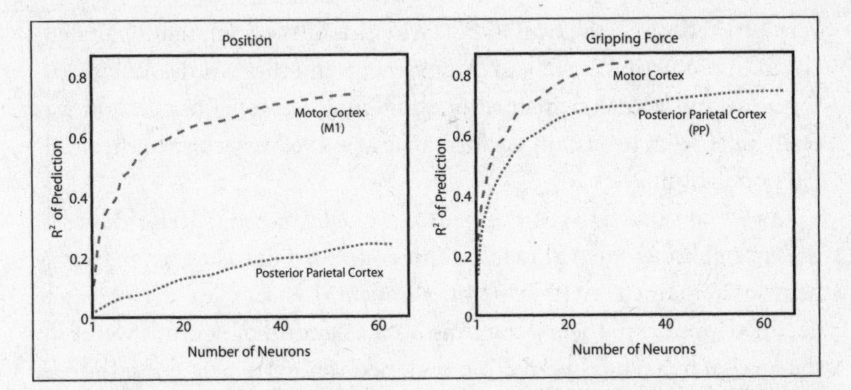

FIGURE 7.2 Gathering motor commands all over the brain. Neuron dropping curves are used to compare the accuracy of prediction of populations of neurons simultaneously recorded in the primary motor cortex (M1, traced line) and the posterior parietal cortex (PP, dotted line) for two parameters (hand position and gripping force). Notice that information about both parameters is available in both cortical regions, but that M1 contains more information, for equal neuronal populations, for hand position. Yet, both M1 and PP yield similar levels of accurate predictions for gripping force, considering neuronal populations of the same size. *(Originally published in J. M. Carmena, M. A. Lebedev, R. E. Crist, J. E. O'Doherty, D. M. Santucci, D. R. Dimitrov, P. G. Patil, C. S. Henriquez, and M.A.L. Nicolelis. "Learning to Control a Brain-Machine Interface for Reaching and Grasping by Primates."* Public Library of Science *1 [2003]: 193–208.)*

the funds to obtain a faster computer, we had to rely on Wessberg's Swedish patience and resourcefulness to get all the curves calculated and plotted.

Despite our computing power limitations, Wessberg persisted and, using every drop of data generated by Belle and Carmen in their efforts to consume as much fruit as possible, produced the first NDCs ever published. These plots, which appeared in *Nature* in 2000, grabbed the attention of the neuroscience community by revealing a couple of very intriguing and provocative findings. First, although neural ensembles from different cortical areas exhibited a clear degree of specialization in predicting one-dimensional arm movements, information about the same motor behaviors could be simultaneously derived from each of the cortical fields from which we recorded neuronal activity in our owl monkeys. Second, as individual neurons were randomly removed from the original recorded population in each cortical area, the NDCs indicated very little effect on the overall performance of the computational algorithm in predicting a particular movement. In other words, despite losing a few individual elements, the remaining neuronal population was resilient enough to sustain performance at a level very close to the original population.

As we continued to remove neurons, one by one, the performance of the ensemble degraded gradually. This continued until just ten to twenty neurons remained. At that point, the neural ensemble's performance declined rapidly, so that by the time only a handful of neurons were left, the level of behavioral prediction obtained was extremely poor. Indeed, whenever only one neuron was available, *no* reliable prediction of movements could be returned by the BMI algorithm.

Since our NDCs were produced by pulling random neurons from the original neuronal sample, these last results implied that, on average, the electrical activity of any single cortical neuron recorded in our experiments could not unambiguously predict the type of movement our monkeys intended to make a few hundred milliseconds into the future, not even if the neuron was located in the primary motor cortex. To generate purposeful arm movement, or to reproduce that movement in an artificial machine, the brains of our two owl monkeys relied on the cooperative work of a population of neurons. This defines the neurophysiological principle of *single neuron insufficiency.*

THE SINGLE NEURON INSUFFICIENCY PRINCIPLE

No matter how well tuned a single neuron may become to a particular parameter, its individual firing rate is insufficient to sustain a particular function or behavior mediated by the cortex. Since the contribution of most single cortical neurons varies significantly from moment to moment, their lack of individual statistical reliability means that no brain-machine interface can be operated consistently, over long periods of time, based on a single neuron's firing rate, and that the basic functional unit of thinking cannot be a single neuron, but instead a population of neurons.

As exciting and demanding as the experiments with Belle and Carmen were, there was a key element that could not be tested during their execution: how would primates react to a feedback signal that informed them in real time of the motor performance of the robots they were controlling with their brain activity?

By the time Aurora had decided to cooperate with us, our experimental apparatus had been upgraded in many different ways to try to address this question. For starters, the development of high-density microelectrode arrays, spearheaded by our resident "manufacturing magician," Gary Lehew, with the help of Jim Meloy, had opened the way to monitoring up to 512 microelectrodes—potentially as many as 2,048 single neurons simultaneously. Such technological capabilities allowed us to observe how Aurora's brain adapted to the demands of learning new tasks and interacting with the BMI, which had also been enhanced to provide feedback information about the performance of a robot arm, placed in another room in our laboratory, in real time.

In a way, Aurora had already been trained to receive feedback from a machine. For example, since the joystick we had given her for playing the video game moved easily, she had to learn how to control her full-of-wild-purpose hand grips and arm movements to produce the desired trajectory for the cursor, using the visual information on the screen. Since her reward depended on how quickly she grabbed the circle target, which randomly popped up on the monitor, she also learned the importance of not wasting time or energy by making erroneous and futile guesses before the target appeared. Instead, she controlled her impulses and sat in an attentive state.

The night we turned on the BMI, removed the joystick, and left her to figure out the new game on her own, she moved her own arm and hand for the first few minutes, as if trying to reach into the screen to grab and squeeze the target. When she stopped making arm movements whatsoever and started to receive her juice reward as before, we realized that she had made the transition from the joystick to BMI control. And because we were continuously monitoring the EMG activity produced by several of Aurora's arm and back muscles, we were able to document the precise moment in which Aurora's brain activity dissociated itself from her body muscles.

As in Eb Fetz's experiments, both the visual feedback and the juice reward worked as powerful reinforcers in this process of dissociating

Aurora's brain and her muscle activity. Curiously, however, we never instructed Aurora to opt for this dissociation, in the way that Fetz and Finocchio had designed their experiments with primates. In fact, as long as Aurora continued to make the cursor cross the target she would receive a drop of fruit juice, whether her muscles were still or contracting. Yet, as soon as she figured out that her arm movements were not needed to win the reward she craved, she took the initiative and refrained from moving her body, only using her mind to accomplish her feat. Nonhuman primates could *voluntarily* opt to dissociate the production of brain activity containing voluntary motor intentions from the translation of these intentions into true muscle action. Indeed, since our EMG records indicated that no muscle contraction occurred in the multiple arm muscles we were monitoring, it appeared that none of the motor neurons in the spinal cord that projected to these muscles were activated either. Somehow, Aurora could almost completely prevent the motor instructions generated in her cortex from being "downloaded" to her spinal cord.

Over a few weeks, Aurora improved her behavioral performance while using the BMI in direct brain-control mode. By the end of this second training period, she could get as many trials correct and drink as much fruit juice as she had done using the joystick to play the game. Moreover, she was able to reduce the time needed to generate the cursor trajectories using the BMI until she reached the same delay observed when she played the game with a joystick. In a mere 250 milliseconds, Aurora's brain activity could be recorded from her brain, rerouted to a central computer, fed into several mathematical models, translated into motor digital commands that could be understood by a machine, transmitted to the robot arm, used to guide a cursor's trajectory, and finally return, most of the time in triumph, to her eyes and to her mouth, where she could taste her victory. About thirty days into her training regime, Aurora realized that she could even use her arms and hand for other important purposes, such as scratching her back or grabbing a passerby neuroscientist, while using her brain alone to play the video game.

Once she had mastered her favorite game, we trained Aurora to perform two other motor tasks. One involved showing on the computer screen a static visual target formed by two concentric circles with different diameters, plotted one inside the other. In this task, the difference between the two circle diameters indicated how much gripping force Aurora had to produce to receive a drop of fruit juice. After learning to solve this puzzle by calibrating the amount of force she used to grip the handle of the joystick, Aurora made a relatively easy transition to employing the BMI to produce the right amount of gripping force just by thinking. Once again, she realized she could fulfill the task requirement without generating overt movements of her own hand.

In a final display of her experimental prowess, Aurora learned to solve a much more elaborate task, one that combined the most arduous elements of her preceding two experiments. As in the first game, Aurora had to guide a computer cursor to intercept a circular target that appeared in a distinct, randomly selected location on the computer screen. However, by the time the cursor reached the target, its shape would morph into the two concentric circles that represented how much gripping force Aurora had to generate in order to hold the target for a few hundred milliseconds. Now, to receive her juice reward Aurora not only had to make the cursor reach the target, but she also had to make the cursor "hold it," applying the correct amount of gripping force. It took a while, but Aurora eventually learned to perform this more complex operation using pure thought, no arm movements required.

Aurora's spectacular performances yielded a ton of neurophysiological data that could be analyzed using the neuron dropping curves. In this case, NDCs could be calculated for each of many different motor parameters simultaneously predicted in real time by our models. We could also take into account each of the six cortical areas from which we recorded single neurons as Aurora mastered her tasks, learned to use the BMI, and dissociated her brain and muscle activity.

The NDCs based on Aurora's data reinforced the observations we had gleaned from our experiments with Belle. In fact, we did not need to dig too deeply into the data to verify that predictive information about Aurora's arm trajectories could be derived from populations of neurons, but not from individual cells, in all six cortical areas surveyed. Using the level of accuracy in predicting the real-time variation of each of the individual motor parameters modeled in our experiments as an initial criterion, we observed that although samples of neurons from different cortical areas displayed different levels of specialization in predicting each of these parameters, neuronal ensembles from every cortical area concurrently carried at least some meaningful information. Take, for example, the NDCs plotted in Figure 7.2. These curves compare the number of neurons needed, per cortical area sampled, to predict hand position or hand gripping force. A much smaller sample of M1 neurons can be used to predict hand position, at much higher levels of accuracy, than an equivalently sized population of neurons recorded in a region of the posterior parietal (PP) cortex. However, when we compare the contribution of the same cortical areas to the real-time prediction of hand gripping force, neuronal samples from the PP cortex generated predictions almost as accurate as M1 ensembles of the same size. If we could have recorded more PP neurons, it is conceivable that larger neuronal ensembles from this region could perform just as well as the M1 ensembles. Not only were the general thoughts about Aurora's arm trajectories generated across vast extensions of her frontal and parietal cortex, but, very likely, many of the cortical neurons participating in one particular motor task could lend their firing activity to the computation of several motor parameters, simultaneously.

That finding led me to develop yet another principle of the relativistic brain, the *neuronal multitasking principle.*

THE NEURONAL MULTITASKING PRINCIPLE

Individual cortical neurons and their probabilistic firing can simultaneously participate in multiple functional neural ensembles. That means that the spikes produced by a single cortical neuron can be utilized by distinct neural ensembles to encode multiple functional and behavioral parameters. Thus, even though in a particular moment in time a single cortical neuron may exhibit sharper tuning to a particular motor or sensory parameter, its spiking may concurrently participate in the encoding of a different parameter, performed by another subset of neurons. The neuronal multitasking principle predicts that the entire cortex is capable of exhibiting cross-modal sensory responses, and that individual neurons are capable of encoding multiple motor, or other higher cognitive, parameters.

There was no sign of strict and precise localization of motor functions in Aurora's brain. Instead, cortical specialization, although evident, was relative; it coexisted with a high level of function sharing. No hint of an encounter with a grandmother neuron seemed apparent either. When neuronal populations from all cortical areas were reduced to a single neuron, none of them individually generated the type of meaningful predictions of Aurora's motor behavior that could drive the BMI to work continuously, trial after trial, with a decent level of accuracy.

The main conclusion that emerged from these experiments was rather straightforward. Within Aurora's brain, the phrenological inheritors to Franz Gall and the single-neuron acolytes were both, at long last, utterly defeated by nature's clever idea of relying on highly distributed populations of neurons to sculpt an animal's behavior.

Neuronal democracy rather than single neuron dictatorship was the slogan written across Aurora's brain.

The next step in our analysis was to compare how individual cortical neurons reacted when Aurora made the delicate transition from playing the video game with the joystick and her own limbs to the brain-control mode of operating a BMI without producing even the slightest contraction of those same muscles. To accomplish this, we analyzed tuning curves that measured how each individual cortical neuron firing correlated with the velocity and direction of movement of Aurora's biological arm and the robot arm at moments in time before, during, and immediately after the execution of the movement. Figure 7.3 depicts three such tuning curves, each built with data derived from a distinct task condition: joystick control mode, brain control mode with concurrent arm movements, and brain control mode with movements of the robot arm without movements of Aurora's own arm.

Based on a vast literature of neuroscientific research published over the last forty years, we expected a large percentage of the recorded neurons to exhibit firing patterns that were somewhat correlated with some aspect of Aurora's arm and hand movements. This was, indeed, what we found. The cortical neurons modulated their electrical activity in a variety

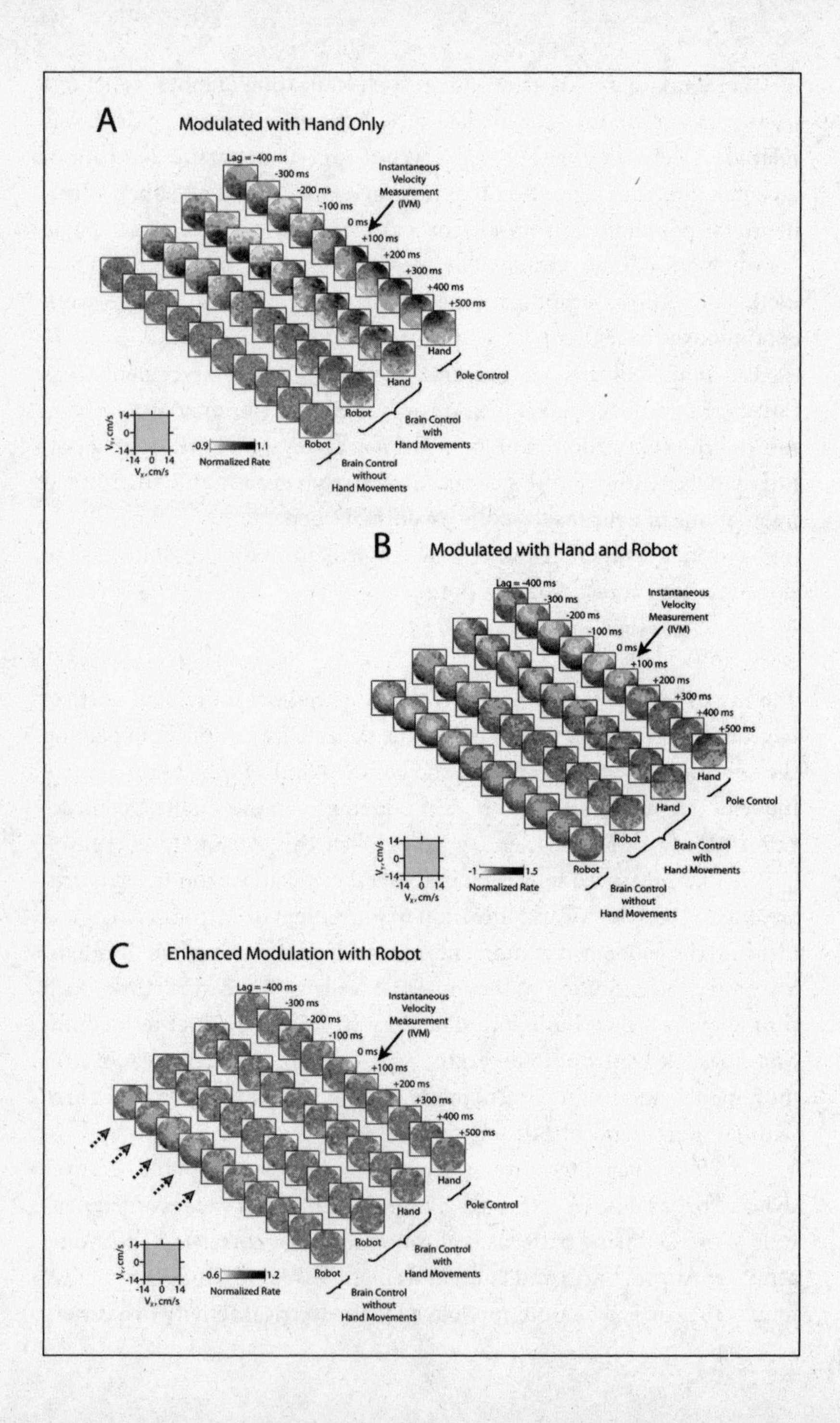
A
Modulated with Hand Only
Lag = -400 ms
-300 ms
-200 ms
-100 ms
0 ms
+100 ms
+200 ms
+300 ms
+400 ms
+500 ms
Instantaneous Velocity Measurement (IVM)
Hand
Pole Control
Hand
Robot
Brain Control with Hand Movements
Robot
Brain Control without Hand Movements
V_y, cm/s
V_x, cm/s
-0.9
1.1
Normalized Rate
B
Modulated with Hand and Robot
Lag = -400 ms
-300 ms
-200 ms
-100 ms
0 ms
+100 ms
+200 ms
+300 ms
+400 ms
+500 ms
Instantaneous Velocity Measurement (IVM)
Hand
Pole Control
Hand
Robot
Brain Control with Hand Movements
Robot
Brain Control without Hand Movements
-1
1.5
Normalized Rate
C
Enhanced Modulation with Robot
Lag = -400 ms
-300 ms
-200 ms
-100 ms
0 ms
+100 ms
+200 ms
+300 ms
+400 ms
+500 ms
Instantaneous Velocity Measurement (IVM)
Hand
Pole Control
Hand
Robot
Brain Control with Hand Movements
Robot
Brain Control without Hand Movements
-0.6
1.2
Normalized Rate

of ways in relation to these movements; a single neuron could fire in anticipation of the onset of these movements or change its firing up and down while the movements were being executed. Such distinct firing patterns could be seen, in more or less different frequencies, in every one of the cortical areas we sampled.

Further analysis of the velocity and direction tuning curves revealed several interesting properties. First, we identified a population of cortical neurons that modulated their firing only when Aurora employed her own limb and hand to generate a movement (see Figure 7.3A). Invariably, these neurons exhibited clear and broad velocity and direction tuning prior to the onset of her arm movements. Often, both the velocity and direction tuning of these cells changed dynamically as the movement unfolded over time. Again, these dynamic changes resembled the flexibility of the spatiotemporal tactile receptive fields of the cortical and

FIGURE 7.3 Fine-tuning of cortical neurons that control the movements of the body and machines. Gray-scaled polar plots of the firing rate of one M1 neuron as a function of both arm velocity (key, top left) for different lags with respect to instant of instantaneous arm velocity measurement (0 ms). Velocity = 0 is at the center of each circle and maximum velocity (14 cms/second) is at the perimeter of the circle. Firing patterns were obtained during different modes of operation (pole control and brain control with and without hand movements), and during the use of different actuators (hand or robot movements, see legends). Each of the circles also encodes the neuron's preferred direction (see scale). Gray shading indicates firing rate (minimum is white, maximum is dark gray). (A) A neighboring neuron that exhibits strong velocity and direction tuning only when the animal is using its hand to play the video game, but not the robot. (B) A single M1 neuron that displays both velocity and direction tuning during both pole and brain control and when the animal is using its hand or only the robot to play the video game. (C) A neighboring M1 neuron that exhibits enhanced velocity and direction (dashed arrows) tuning only when the monkey is preparing to use its brain activity to move the robot arm, but not when it moves its own biological arm. *(Originally published in M. A. Lebedev, J. M. Carmena, J. E. O'Doherty, M. Zacksenhouse, C. S. Henriquez, J. Principe, and M.A.L. Nicolelis. "Cortical Ensemble Adaptation to Represent Velocity of an Artificial Actuator Controlled by a Brain Machine Interface."* Journal of Neuroscience *25 [2005]: 4681–93, with permission.)*

subcortical neurons recorded in our whisking rats a decade earlier. Second, once Aurora stopped making movements with her own arm, this group of cortical neurons stopped firing altogether. No action potential spikes were created. As a consequence, these neurons showed no velocity or direction tuning to the movements of the robot arm, despite the fact that these movements were actually under the control of Aurora's brain (see Fig. 7.3A).

But this highly specific covariance of neuronal firing with biological arm movements was only one of the patterns that we were able to see within the electrical activity in Aurora's cortex. Another sizable subset of her cortical neurons displayed velocity as well as direction tuning in response to movements of both her biological arm and the robot arm, even when Aurora was manipulating the BMI without making any muscle contractions (see Fig. 7.3B). Sometimes the velocity and direction tuning changed as Aurora made the transition from moving her own arm to directly controlling the robot limb with her brain alone. But in many cases, the velocity and direction tuning of a single cortical neuron remained the same or very similar in both conditions. Clearly, some of Aurora's neurons could fire in the absence of any overt movement in her arm, as Eb Fetz observed among his monkeys. Moreover, our experiments showed that these cortical neurons could also sustain their original tuning characteristics when Aurora needed to control the movements of the robot arm with her brain alone.

Easily, this latter observation could have emerged as the most relevant finding in all of our experiments with Aurora. It turned out, however, that it was soon overshadowed by the identification of a third class of cortical neurons whose physiological behavior stunned us beyond belief. In a total departure from anything I had thus far seen or imagined in my career as a neurophysiologist, this subset of cortical neurons did not fire at all when Aurora moved her own arms or hands, meaning that they did not exhibit any velocity or direction tuning when Aurora was controlling the joystick with her hand. Yet, the moment Aurora started to employ her brain activity alone to drive the BMI and control the robot arm, these neurons started to fire like crazy and soon acquired gorgeous tuning for the velocity and direction of movement of the robot arm (see Fig. 7.3C, dashed arrows). Curiously, these neurons were found throughout the cortex, even in M1, sitting next to other neurons that fired only when Aurora moved her own arm. So much for the existence of order and structure in the primate primary motor cortex. Essentially, we interpreted this observation as if a fraction of Aurora's cortical neurons sat quietly in her sensory motor cortex until the moment they could seamlessly assimilate a tool or, in this case, a robot arm as part of her brain's internal simulation of her body image.

For those of us committed to developing BMIs, Aurora's results sounded like the final movement of Beethoven's Ninth Symphony: pure hope and joy. The reason for our optimism was simple. If Aurora could dissociate the production of motor activity in her brain from the contraction of her body muscles, it was very likely that patients suffering from severe lesions of the spinal cord or a devastating peripheral neurodegenerative disorder, which render them paralyzed without affecting the rest of their brains, could learn to use their cortical activity to control the movements of a neuroprosthetic device designed to restore full body mobility.

Although few noticed at the time, Aurora's valiant efforts in these classic experiments created a unique convergence of ideas, dreams, and semi-lost causes that had been stalking the marginal, often neglected corridors of mainstream neuroscience for nearly two hundred years. By fully embracing Thomas Young's distributed coding insight, Karl Lashley and Donald Hebb's concept of regional equipotentiality and cell assemblies, John Lilly's obsession to see and hear as many neurons flashing simultaneously as possible, Eb Fetz's pioneering use of feedback, and Edward Schmidt's dream of neuroprosthetic devices, the Aurora experiments showed that it was possible to redefine the model of self that only a primate brain can create.

In the proverbial great relay race of science, it seemed that the hope of turning the final corner was finally within arm's reach.

8

A MIND'S VOYAGE AROUND THE REAL WORLD

Melding brains and machines had, for decades, appeared to be a far-fetched dream or, at best, the stuff of science fiction. With the publication of our research with Belle and Aurora, however, brain-machine interfaces had crossed the threshold into the rarified halls of "real" science. *Scientific American* and the *MIT Technology Review* both highlighted the development of BMIs. A special issue of *Nature* in 2001 was dedicated to reviewing state-of-the-art research and technologies with the potential to influence the future of society. In the article I wrote for that issue, I revealed for the first time a systems-engineering diagram describing the components of our closed-loop control BMI and the steps necessary to transform such a device into a neuroprosthetic device. With all of this attention, neuroscience laboratories around the world started to shift their research and resources to the field.

At that time, the argument about BMIs was dominated by the discussion over whether it was better to employ noninvasive methods, like the scalp EEG, or more invasive chronic brain implants of microelectrode arrays, which, as we had seen with Belle and Aurora, had created a BMI that could effectively use voluntary brain activity to control a robotic arm.

Those working with the EEG-based brain-computer interface (BCI) contended that because EEG signals could be obtained without the need for any invasion of brain tissue, they offered the best balance between clinical risk and clinical benefit. Invariably, defenders of the noninvasive approach pointed to Niels Birbaumer's success in getting "locked in" patients to communicate with the external world through the first clinically relevant application of an EEG-based BCI. A few years later, several groups, including one led by my good friend Klaus Müller, a computer scientist at the Technical University of Berlin, expanded on Birbaumer's BCI concept and developed applications in which healthy subjects could utilize their EEG activity to play simple video games. More recently, EEG-based BCIs have allowed severely paralyzed patients to steer a wheelchair.

Despite these extremely useful applications, EEG-based BCIs had limitations. Because EEGs averaged the synaptic and firing activity of tens of thousands of cortical neurons, the signals fed into BCIs lacked the spatial resolution needed for a prosthetic device to be able to mimic natural limb function. Essentially, since the EEG signal carries very little specific neuronal information, BCIs based on these signals can handle just a couple of bits of information. The debate finally reached a truce when the two camps realized recently that BMIs could be built by taking the best approaches from noninvasive and invasive techniques. In addition, in patients suffering from spinal cord injuries, BMIs could one day be used together with other revolutionary therapies like stem cells to restore full body mobility.

In late 2002, my lab at Duke obtained preliminary human intraoperative data that demonstrated the viability of using the BMI approach in clinical applications. We had decided to get as practical as possible, enlisting Dennis Turner, a neurosurgeon at Duke's medical school, along with two of the residents studying under him, Dragan Dimitrov and Parag Patil. Our chief technology developer, Gary Lehew, worked with the neurosurgical team to adapt a commercially available microelectrode to test the concept of a BMI during a standard surgical procedure for patients suffering from advanced symptoms of Parkinson's disease, a degenerative disorder in which neurons rich in the neurotransmitter dopamine, a key chemical utilized by the motor system to initiate and produce smooth voluntary movements, die over time. The surgery involved implanting a deep brain stimulator, an electrode about the size of a heart pacemaker that is used to block the abnormal nerve signals that cause tremors, stiffness, walking difficulties, and other Parkinson's symptoms. Because it does not halt the degeneration of dopaminergic neurons, deep brain stimulation (DBS) does not offer a cure for Parkinson's disease; it is simply the most effective treatment available for patients who no longer respond to pharmacological dopamine-replacement therapy.

Normally, Turner and his team gain access to the brain by opening the skull and the meninges while the Parkinson's patient remains conscious. This is possible because direct manipulation of brain tissue does not generate any sensation of pain; the tissue does not contain pain receptors, or nociceptors. (Thus, even though the brain continuously signals the pain of the body that houses it, it never senses its own sorrows.) Turner next identifies a very small spot, deep in the brain, in which continuous electrical stimulation, delivered through the chronically implanted electrode, will ameliorate some of the patient's most serious motor symptoms.

The most crucial step is the correct placement of the stimulator in the patient's brain. During the surgery, in addition to asking the patient about the effects of the stimulating electrode, Turner guides its placement by continuously monitoring the electrical activity of the neurons he encounters as he penetrates the brain with a probe. Usually, this process is carried out using the classical single-microelectrode technique in which one neuron (or even multineuronal activity) is recorded serially as the tip of the electrode is slowly lowered into the cortex. In the event that Turner cannot find a usable spot during the first brain penetration, the electrode must be withdrawn and the process is repeated with a new brain penetration. Usually more than one brain penetration is made before the patient confirms that the stimulation has quelled his or her tremors and other Parkinson's symptoms.

Lehew helped Turner and his colleagues improve on this routine by adapting a commercially available probe that had been approved for neurosurgical procedures to record many more spots simultaneously. In fact, rather than being limited to one spot at each depth of the brain at which he had parked his probe, Turner could now monitor thirty-two spots. Lehew had made this possible by bundling together thirty-two microwires into a single recording probe that could be lowered into the brain with a guiding tube. This microwire bundle immediately reduced the time necessary to complete the procedure. Indeed, in most cases the neurosurgeons could find a spot that worked in a single penetration.

After we were able to demonstrate the clear clinical benefit of Lehew's microwire innovation, we were granted permission by the Duke Institutional Review Board—and, as important, eleven patients—to test a simple version of the BMI we had used with Aurora during these procedures to implant a deep brain stimulator in Parkinson's patients. Since the surgeries were now faster, we were given a limited amount of time to explore the efficacy of our apparatus. We therefore decided to train the patients to play a very simple video game the day before surgery was scheduled. In this game, the person used one of his or her hands to squeeze a rubber ball. By applying different amounts of gripping force to the ball, the patient could move a computer cursor back and forth on a track line on a monitor. The aim: to get the cursor to hit a rectangular target that popped up at a location along the track. It took each patient just a few minutes of practice to master this task.

During the surgical procedure, the same patients were asked to play the game for about five minutes while their brain activity was recorded by the thirty-two-microwire probe as it slowly penetrated the brain. Unlike in our experiments with Aurora, in these intraoperative recordings we could simultaneously monitor at most fifty neurons. Moreover, since patients with severe Parkinson's symptoms often tire rapidly during the DBS procedure, we tried to minimize their fatigue by limiting our study to the collection of data needed to verify whether the BMI would become capable of predicting the cursor's trajectory using the patient's brain activity alone. In the end, that meant we recorded just ten minutes of data in each of the patients—five minutes dedicated to training BMI and five minutes to test it.

Even with this small window for recording, the results were resounding. The BMI algorithms pioneered by Aurora and her monkey friends also worked well when they were fed with human brain activity. Johan Wessberg's mathematical models could predict simple hand movements for humans, too. More important, as the number of neurons employed to feed the BMI dropped from fifty neurons to thirty or twenty, the accuracy of the hand-movement predictions fell precipitously, to a point at which they became useless. It was true that our study had recorded activity from subcortical motor structures, such as the thalamus and the subthalamic nucleus, rather than the M1 cortex. But there was no indication that very small samples of neurons in M1 would behave significantly different from those in these structures.

After we successfully completed the experiments with Aurora, it was essential to explain what it means to say that BMIs can liberate the brain from the physical constraints imposed by the body that lodges it. One way to do this is to look for clinical applications of our BMI experiments.

Before we could move safely and confidently to more clinically oriented work, however, the laboratory team at Duke needed to explore further animal experiments. By the fall of 2006, we had made some of our first steps into the remaining uncharted territory. Thanks to the superb work of my graduate student Nathan Fitzsimmons, who also illustrated this book, we had learned that direct cortical microstimulation, using the same microwire arrays we used to record brain activity, could potentially deliver feedback messages directly to the brain of owl monkeys like Belle. Yet, such an approach still needed to be implemented in rhesus monkeys, during the operation of a closed-loop BMI, before we could be sure of this outcome. Since these demanding experiments would require extensive behavioral training and a series of control studies, we decided to move forward with another project based on a fundamental observation derived from our work with Belle and Aurora.

We had learned from Aurora that the utilization of a BMI could allow a subject to change simultaneously the normal scale of three physical parameters involved in the generation of primate motor behaviors: space, force, and time. By directly controlling the movements of a robot arm located far away from her body, Aurora had increased dramatically the spatial reach of her voluntary motor intentions. Since the robot was capable of generating stronger forces than those normally produced by her own arm, Aurora also scaled up the force that resulted from her motor thinking. Finally, through direct control of our upper-limb BMI, Aurora could generate movements in the robot arm a bit faster than the normal neurobiological process through which she moved her own arms, amping up the time scale, as well.

By the early months of 2007, the main questions we were debating in our laboratory gatherings, usually over some generous servings of Greek-Lebanese food at George's Garage, our favorite restaurant in Durham, hovered around two main issues. The first related to whether BMIs could be used to produce more than upper-limb movements. In other words, could BMIs ever serve as a platform for restoring all sorts of motor behaviors? The second issue dealt with how far we could possibly push the scaling of space, force, and time involved when a subject operates a BMI.

As things in science usually go, the answer we were looking for did not present itself in the circumspect and well-behaved environment that usually dominates a typical systems neurophysiology lab. In this case, the hard-sought epiphany came to me in the middle of an intensely contested soccer match, played (and barely won) by my beloved Sociedade Esportiva Palmeiras, in the Palestra Itália stadium in São Paulo where I have religiously attended some historic, as well as some not so memorable, games since my newborn days. That Sunday afternoon, as I watched in distress as the Palmeiras players missed one easy scoring opportunity after another, the desperate crowd repeatedly shouted them down with one of the worst epithets available to them: *pernas de pau*—literally, wooden legs. It suddenly dawned on me in which new direction we should steer our BMI research program. Back in my hotel room, after a brief celebration of Palmeiras's razor-thin victory, the only thing on my mind was whether we could possibly make our rhesus monkeys learn to operate a BMI designed for reproducing bipedal locomotion in a remotely located robot using only cortical neuronal activity as the driving signal.

Without much noise, while I was on a sabbatical at the Swiss Institute of Technology of Lausanne during most of 2006 and 2007, one of my postdoctoral fellows, Andrew Tate, had been training rats to walk on a treadmill in an attempt to measure the concurrent patterns of neuronal electrical activity produced by both cortical and subcortical motor structures known to be involved in the production of locomotion. Traditionally, neurophysiologists interested in the neural mechanisms of locomotion had primarily focused their investigations on subcortical pools of neurons that exhibited a certain rhythmic firing—the so-called central pattern generators—related to the production of a quadrupedal gait cycle. Most of these studies, conducted in cats, had found central pattern generators in the spinal cord and brain stem. It seemed that the existence of central pattern generators at the spinal cord level explained why cats were able to sustain a quadrupedal locomotion pattern even after the spinal cord had been severed from the rest of the animal's brain: place a cat with a severed spinal cord on a moving treadmill, and the cat can keep up. In primate evolution, most of these central pattern generators had moved into the brain, which is why a spinal cord injury or lesion induces an irreversible paralysis of the body musculature below the level of damage. Even if the rest of the body is suspended, a human patient with such an injury will not be able to keep up with the treadmill's pace.

Because of the long tradition of focusing on subcortical central pattern generators in experimental animals, as well as the difficulties involved in performing chronic cortical recordings in freely ambulating animals, the potential role of the motor cortex in controlling the gait cycle in nonhuman primates had been pigeonholed as a stubborn scientific challenge. To our surprise, however, Andrew Tate's rat studies revealed that cortical neurons located in both the M1 and S1 cortices did modulate their firing rate when the animals walked at normal speed on a moving treadmill.

This revelation spurred us to take the risk and invest in building a completely new apparatus that would allow us to test a BMI for creating locomotion. To begin, though, we needed to demonstrate that rhesus monkeys could actually walk bipedally on a treadmill. Then, we had to demonstrate that we could record the electrical activity simultaneously produced by a few hundred monkey cortical neurons as it walked. No study that we were aware of had reported that rhesus monkeys could perform such a behavioral trick, and none had recorded chronic cortical activity in freely walking monkeys.

Luckily, an answer to one of our problems came from a most unusual source when Misha Lebedev, my long-time collaborator and the newly appointed director of our primate laboratory, dug out some early-twentieth-century reports describing how Russian circuses had trained rhesus monkeys to "walk bipedally on a platform." Bingo! The secret was in providing enough support to the upper torso of the animal so that it felt secure enough to stand on its hind limbs and start walking.

We assigned this impossible job to our technology whiz Gary Lehew.

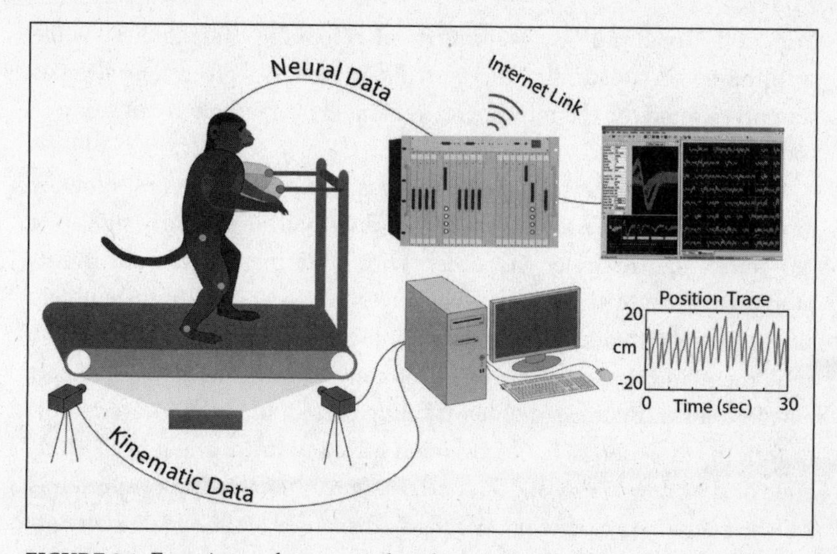

FIGURE 8.1 Experimental setup employed to create a brain-machine interface for locomotion. *(Originally published in N. Fitzsimmons, M. A. Lebedev, I. Peikon, and M.A.L. Nicolelis, "Extracting Kinematic Parameters for Monkey Bipedal Walking from Cortical Neuronal Ensemble Activity."* Frontiers in Integrative Neuroscience *3 [2009]: 1–19.)*

By now used to our absurd demands, Lehew did not take long to figure out how to build a "walking-monkey" setup, consisting of a hydraulic treadmill outfitted with a Plexiglas support for the monkey's upper body, allowing it to walk peacefully along with the direction, speed, and slope at which the treadmill was set (see Fig. 8.1). By using a hydraulic mechanism instead of an electric motor, he conveniently eliminated a major potential source of electrical noise from our neuronal recordings. The treadmill was placed in a soundproof, shielded room, further ensuring that the monkey would be protected from any distraction during its daily strolls.

Lehew also designed a clever tethering system to hold some of our recording hardware right above our subject's head. This adaptation ensured that the many cables connecting the implanted microwire arrays to the preamplifiers did not get tangled by the walking monkey. To allow the monkey to receive feedback from the machine it was controlling via the BMI, he installed a system to project video feeds on the wall in front of the treadmill. Any walking monkey looking straight ahead while strolling on the treadmill—the most likely scenario, given the Plexiglas support—would see the movements generated by a remote robotic device controlled by the BMI.

As soon as this new apparatus was ready, we chose the first monkey to test it. Like her predecessor Aurora, Idoya had shown early signs of a "go-getter" nature in her encounters with other members of our rhesus monkey colony. Although she had never ever seen anything remotely close to Lehew's walking-monkey setup, she was not intimidated by the equipment and, in just a few weeks of daily training, became an expert bipedal walker. By the end of her training period, Idoya had learned how to shift the direction of her locomotion back and forth, and how to accelerate or slow down her walking speed as soon as the treadmill's speed was adjusted. As long as her fruit reward was dispensed at the end of a few correct steps, she had no problem walking for an hour or more per day.

At this stage, we implanted several microwire arrays in some cortical areas of Idoya's brain. A few days after the surgery, she was back walking, like any addicted primate jogger. The difference now was that our implants were yielding recordings from hundreds of gorgeous neurons, which showed a clear modulation in their firing rate as Idoya walked along, unconcerned. These initial recordings revealed that single neurons in both the S1 and M1 cortices tended to fire prior to the initiation of each step cycle. When combined into a population and plugged into the linear multivariate models employed in our upper-limb BMI, these cortical neural ensembles generated highly accurate, real-time predictions of Idoya's strides. That brought us to the next major bottleneck in our attempt to generate true locomotion patterns out of a primate's raw brain activity: which artificial walking device could possibly utilize in real time the type of brain-derived control signals that we could record from our primate's brain? That was a tough one. To make sure that everyone grasped the implications of our experiments, we needed to demonstrate walking in some sort of humanoid robot.

As it happened, a few years earlier I had actually met a humanoid robot who fit the bill. That robot resided in the lab of Gordon Cheng, the founder of the Department of Humanoid Robotics and Computational Neuroscience at the Advanced Telecommunications Research Institute (ATR) in Kyoto, Japan. In 2005, I had met Cheng and ATR's director, Mitsuo Kawato, during a visit to see a still-under-construction robot named Computational Brain, Model 1, or CB-1 for short.

Designed by Cheng and built by the American company Sarcos, CB-1 looked pretty humanoid, with its two legs and arms, despite being powered by a hydraulic system. The robot was primarily going to be used to study the possibility of reproducing realistic, humanlike motor behaviors, including locomotion. Using a previous generation of the same robot family, Kawato had already achieved arm and hand movements that were highly coordinated and precisely aimed—a wonderful triumph. They had even been programmed to play Ping-Pong and execute a shuffling version of the steps of several traditional Japanese folk dances. I was sure I had the right roboticist and the right robot to marry with our walking monkey's brain. So I called Cheng up one day to make a proposal—one that I had not even mentioned in a recent grant proposal for fear of being ridiculed by the reviewing panel.

"So, my friend," I began tentatively. "We have figured out that one of our rhesus monkeys, Idoya, can walk bipedally on a treadmill, and that the collective neuronal activity we record from her cortical areas can predict her gait cycle. That means we have built a BMI that in theory could allow Idoya's brain activity to drive your CB-1 robot, in real time, to be the first humanoid robot to ever walk under the control of a primate brain."

That was the easy part.

"We still need to figure out how to get the brain data to Japan quickly enough so that the robot can walk as fast as it takes Idoya's legs to respond to her brain. We also need to figure out how to return a video feedback signal from the robot's legs to Idoya so that she is aware of what is going on at every moment. As you know, both the transmission of the brain signals from the United States to Japan and the return of the visual feedback from Japan to the United States have to occur within the same time delay as the animal's reaction time."

That was the hard part. As we had seen with Aurora, we had just a few hundred milliseconds to play with. Longer than that, and the machine and the brain would fall out of sync, and the BMI would fail.

"Sure. Let's do it. I like the sound of the idea. Believe me, I can get the signals back and forth between Durham and Kyoto faster than 250 milliseconds."

"Are you sure?"

"Yeah. Let's start tomorrow." That is why I like Cheng so much; he is always ready to take on the almost inconceivable.

Before I was up the next morning and could tell the team in the Duke bunker about this good news, Cheng was busy coding a way to get around the massive Internet traffic jam that confounded us in order to hit his 250-millisecond mark. For the next few months, in addition to his own lab projects, he worked on the problem every night.

The data transmission speed wasn't the only hurdle placed before our transcontinental human-robot team. For instance, Ian Peikon, a brilliant undergraduate engineering student, almost single-handedly designed and implemented a completely new system for measuring the continuous three-dimensional spatial position of Idoya's leg joints. Peikon's kinematic recording system involved placing on the ceiling above the treadmill a number of video cameras, each of which measured the light reflected by a layer of fluorescent paint applied to Idoya's right hip, knee, and ankle. This tactic generated a continuous stream of precise information about the position, velocity, and acceleration of Idoya's leg joints as she walked on the treadmill. We then asked Cheng to add to his list of projects a robotics interface for utilizing the brain-derived data that predicted this kinematic data.

Peikon's kinematic data and the concurrent recordings of the electrical activity of a couple of hundred of Idoya's cortical neurons were then fed into the same computational algorithms that we had employed in Belle's and Aurora's experiments. When we analyzed Idoya's brain activity as she walked at constant velocity on a treadmill, we found that steps which appeared identical from a kinematic point of view were preceded by distinct, fine-grain spatiotemporal patterns of neural ensemble firing—measurements that were made with just a couple of milliseconds of temporal resolution. If we integrated the activity of this small number of neurons at a time interval on the order of one hundred milliseconds, however, the variability of these patterns of cortical activity was reduced significantly. This implied that the 250-millisecond interval could be successfully employed in a BMI.

It also suggested something more profound. Since billions of neurons are available to participate in producing a solution to any given problem at any given moment in time, the brain has the luxury of being able to recruit a different combination of neurons each time it desires to produce a motor behavior. Indeed, I propose that throughout our lives, no matter how many times we repeat the same motor behavior, the fine-grain spatiotemporal neuronal pattern that carries this particular voluntary motor desire will never be exactly the same. This finding, which resembles ideas originally put forward by Karl Lashley and Donald Hebb, is encapsulated in the *neural degeneracy principle*, a term coined by the American Nobel laureate Gerald Edelman, who compared this strategy to the degeneracy, or redundancy, observed in the genetic code.

THE NEURAL DEGENERACY PRINCIPLE

A particular brain outcome—whether a motor action, a perceptual experience, even complicated behaviors produced by the brain such as singing or solving an equation—can be generated by a very large variety of distinct neuronal spatiotemporal patterns of activity.

In the genetic analog, distinct nucleotide triplets, known as a codon, in a messenger RNA molecule can call on a given amino acid to join a polypeptide chain in a ribosome, the granules within cells where proteins are synthesized. Notice that there is no ambiguity in the genetic code since there is never doubt about which amino acid a particular codon calls for. In the brain, however, multiple neuronal solutions may exist for encoding any goal-oriented behavior the brain needs to produce.

With this insight, by the summer of 2007 we had discovered that by feeding the brain activity that Idoya experienced as she walked on the treadmill at fixed and variable speeds into twenty-one linear mathematical models running in parallel, we could generate the necessary continuous stream of motor commands that would make CB-1 walk like a monkey (see Fig. 8.2). Later that summer, Cheng proved that he was up to the

250-millisecond challenge. He had established a special Internet connection between our two laboratories, a masterpiece of technological invention that circumvented the idiosyncrasies of the Duke and ATR firewalls, outmaneuvered delays caused by the servers and hubs around the world that would route the data between the two campuses, and documented the efficiency of both directions of data transmission at the same time.

The usual protocol at this point would have been to run the experiment, collect the data, publish a paper in some specialized scientific journal, and then disclose the findings to the public, about two years down the road. But we felt that the mind-bonding between Idoya and CB-1 deserved a streamlined process, so we resolved to accelerate the peer review system a bit and run the experiment with the *New York Times* sitting on the sidelines in Durham and Kyoto. The benefits of getting our results disseminated broadly far exceeded the risks of the experiment not working or upsetting our more traditional peers. If we could show that CB-1 was walking under the control of Idoya's brain, it would help bolster the argument that it was within the reach of science to build a BMI, perhaps within a decade, that would allow a person suffering from a devastating neurological disorder to walk again.

On a colder than average January morning in 2008, we decided to stage what I had been calling our "little moon walk." Idoya, the primate poised to make this next great leap, was escorted into the room housing her familiar "walking-monkey" setup. She may have noticed that those assembled in the lab looked conspicuously graver than they had been at her previous walking sessions. As with most major technological endeavors, the Duke team turned to our checklists to prepare for countdown. The level of worrying in the room could be calibrated against the number of times one of our graduate students inspected the many computers involved in sending and receiving data from Idoya's brain and the Kyoto lab, as well as the number of people who asked me if I was sure we had enough raisins and Cheerios—Idoya's preferred food rewards—in stock.

With photographers documenting every moment, Idoya was gently

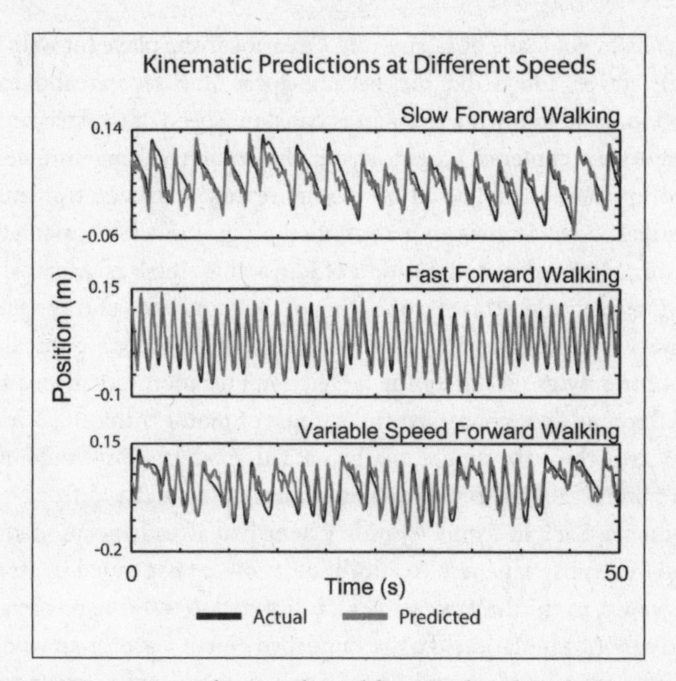

FIGURE 8.2 Kinematic predictions, derived from combined raw brain activity, for different types of bipedal locomotion behaviors: slow forward walking (top shelf), fast forward walking (middle shelf), and variable speed (lower shelf). Black trace depicts actual position of Idoya's leg whereas the gray traces yield real-time predictions of this kinematic parameter obtained from her brain activity alone. *(Originally published in N. Fitzsimmons, M. A. Lebedev, I. Peikon, and M.A.L. Nicolelis, "Extracting Kinematic Parameters for Monkey Bipedal Walking from Cortical Neuronal Ensemble Activity."* Frontiers in Integrative Neuroscience *3 [2009]: 1–19.)*

placed on the treadmill. Before she was allowed to walk free, we linked the plastic connectors, which held the microwire arrays we had implanted in her brain several months earlier, to our hanging preamplifiers. On the projection screen, she could see the sharp, colorful image of CB-1's humanoid legs, which had been magnified a few times so that Idoya's entire visual field was primarily occupied by them.

When the lights were dimmed, the fluorescent paint glowed green at Idoya's hip, knee, and ankle. The conditions seemed to urge Idoya into action. She realized it was time to play her favorite game and that she was likely to get loads of raisins and Cheerios if she played it well. With that incentive, Idoya did not hesitate for a millisecond and eagerly started to walk, following the gentle, constant speed of the treadmill.

The video cameras spread across the room's ceiling immediately picked up the light reflected by the fluorescent markers, transmitting individual video frames to a computer programmed to calculate the three-dimensional spatial position of Idoya's moving legs. Meanwhile, a continuous parallel stream, consisting of the thousands of tiny sparks of voluntary brain electricity made by hundreds of Idoya's cortical neurons, promptly lit up one of the large computer monitors placed in the control room. As we observed this sample of motor thinking, our computers crunched the linear mathematical models that would match Idoya's brain activity to the kinematic parameters derived from her leg movements. Back in Kyoto, Gordon Cheng had already positioned CB-1 on top of a spiffy Japanese treadmill of its own. Suspended by its back, CB-1 would, in this first experiment, be limited to walking on air, a simplified robotic emulation of a distinguished North Carolinian who used to do this for a living, playing basketball, first for Duke's eternal rival, the University of North Carolina, and later on for the NBA (see Fig. 8.3).

As usual, the training of the mathematical models continued for a few minutes until the linear regression coefficients started to converge and stabilize. When the optimal set of regression coefficients was identified, we turned our attention to another computer screen, where Idoya's brain activity appeared to already be predicting her walking movements quite well. The two lines, one red and the other white, moved closer and closer together, until they almost completely overlapped. It was time to start sending the brain-derived predictions to Kyoto.

A few seconds passed while Cheng turned on the control system responsible on his end for streaming the BMI's signals to CB-1. Oblivious to the tension that gripped the humans around her, Idoya continued to maintain her stride without missing a step. Suddenly, the video frames projected on the wall acquired an almost human sense of purpose. A little twelve-pound, thirty-two-inch-tall rhesus monkey was using her mind's electricity to control the primatelike baby steps of a two-hundred-pound, five-foot-tall humanoid robot on the other side of the Earth. It was impossible not to deem it "one small step for a robot, one giant leap for primates."

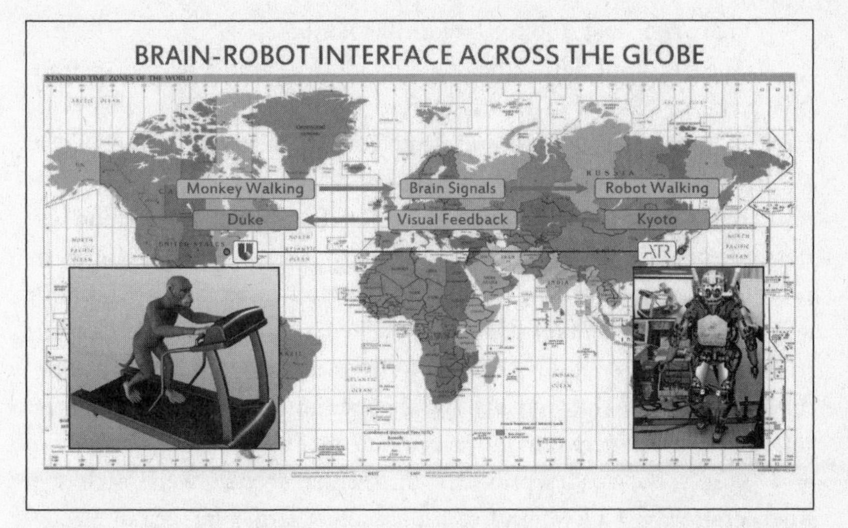

FIGURE 8.3 Idoya and CB-1's great leap across the globe. General schematic representation of the experiment that allowed a monkey on the U.S. East Coast to use its brain activity to control the leg movements of a humanoid robot (CB-1) in Kyoto, Japan, while receiving visual feedback back in Durham, North Carolina, from the robot locomotion. *(Courtesy of Dr. Miguel Nicolelis.)*

Aurora would be proud of her little cousin. After all, in just these few steps, Idoya had pushed the scaling of space and force of brain-machine interfaces to the edge of what was theoretically conceivable, but not yet demonstrated, at the time.

But that was not all. As Cheng quickly reminded me over the phone, we still needed to measure the total time delay elapsing between sending brain-derived signals from Durham to Kyoto and returning video images from Kyoto back to Durham. "It's 230 milliseconds," he reported. "I told you we would do it below 250 milliseconds!" After months of exhausting work, Cheng could not hide his satisfaction. Together with Fitzsimmons, Lebedev, Peikon, and the rest of our team, he had managed to establish a direct, bidirectional functional link between a primate brain and a pair of robot legs that accomplished its task a few tens of milliseconds faster than the time it took for the electrical activity generated in Idoya's brain to trigger a muscle contraction in her own leg. With so much to celebrate, we decided we could take yet another risk.

"On my sign, just stop the treadmill. Okay: stop it."

As the treadmill slowed to a stop and Idoya assumed a typical immobile posture, all eyes in Durham were glued to the monitor showing CB-1 in Kyoto. Idoya, too, seemed intrigued as she continued to stare at the images projected in front of her.

Perhaps she wanted to prove something. For, as long as we stood witness, the only thing we could see was that son of a gun CB-1 robot, only recently promoted to a bipedal walker, continuing to take its rhythmic strides floating in midair. CB-1 just walked and walked, following the instructions that continued to flow, uninterrupted, from Idoya's brain.

Back in Durham, nothing was happening, at least in movement terms. All of us, including Idoya, remained still and silent, admiring in awe those robot strides projected on the screen before us, each step finely crafted, just a few hundred milliseconds earlier, by the raw electrical breath of life that emerged, almost as a divine gift, from a rightly joyful, because it was now a liberated, primate brain.

9

THE MAN WHOSE BODY WAS A PLANE

Starting a few million years ago, when the first hominids began to roam through the valleys of East and North Africa, the brains of some of our Homo genus ancestors underwent a series of morphological and physiological transformations that resulted in the emergence of a cascade of mental processes and behaviors never before witnessed in the animal kingdom. Among other changes, this complex brain remodeling included an enormous differential growth of the frontal and parietal cortical lobes and the many parallel and reciprocal neuronal pathways that link these regions to one another and to a variety of subcortical structures. This huge evolutionary expansion of the frontoparietal circuitry yielded a suite of unique neurophysiological adaptations culminating in new perceptual, motor, and cognitive behavioral skills that contribute to the unique attributes of what we commonly define as human nature.

The ability to produce and comprehend oral language surged during this evolutionary quantum leap in brain architecture. Since many interesting scholarly articles and books have been written on the subject of language and its role in subsequent human evolution, here I will focus instead on two other emergent concurrent adaptations whose inception in the primate brain's cognitive tool kit has been equally decisive, if not essential, in the unfolding of human evolution. The first of these involves the ability of our species and its ancestors to become the ultimate master toolmakers on Earth. Indeed, the presence of many artifacts in close proximity to the early hominid fossil remains found at Olduvai Gorge, in Tanzania, was so conspicuous that the great paleontologist Louis Leakey named the species he unearthed *Homo habilis*—literally, "handy man." The mental abilities required for toolmaking stand as one of our most amazing evolutionary riddles.

Curiously, a second, and perhaps even more revolutionary, behavioral adaptation arising from the burgeoning of the frontoparietal circuitry has attracted less attention in the neuroscience community. Such an attribute ensures that besides being the ultimate toolmakers in the evolutionary history of our planet, the human species has acquired the capability of seamlessly incorporating the artifacts we fabricate into our sense of self, as true expansions of the elaborate and intimate simulated model of our bodies generated by our brain. Although this may sound more incredible than feeling the presence of a phantom limb or having an out-of-body experience, a range of psychophysical, imaging, and neurophysiological experimental evidence for the phenomenon of tool incorporation has been simulated in both human and nonhuman primates. In this chapter I will review this rather stunning neurophysiological evidence. But first I would like to tell a story that illustrates the profound relationship that we, as a species, have established with artificial tools. Created first in our minds, as mere dreams, and soon after translated into tangible mechanical, electronic, and, more recently, computational and virtual artifacts, for the past six million years tools built by the human brain have helped extend our collective reach to the limits of our imagination, allowing, among other things, our terrestrial bodies to conquer the very skies that endowed us with the basic elements of life.

In the daily routine of Parisians during the opening months of the twentieth century, the publicized intention of a young expatriate scientist to test his latest invention was hardly reason for special notice. After all, Paris was at the time one of the leading world centers in science, home to world-renowned physicists, mathematicians, engineers, and inventors. Just a dozen years earlier, Gustave Eiffel had erected the tallest man-made structure in the world, an emblematic demonstration of France's engineering power and ingenuity. Yet, for the crowd of onlookers who, on the windy Saturday afternoon of October 19, 1901, opted to stroll absentmindedly through the immaculate boulevards and manicured gardens of their enchanting city, history had reserved quite an unforgettable treat.

That frigid fall day, Alberto Santos-Dumont, a diminutive but impeccably dressed Brazilian man, defied the normal protocol of discovery by doing something that was as foreign to the scientific community then as it is today: he performed one of his most daring experiments in public, in broad daylight, so that all of Paris could verify whether he had succeeded or failed in his latest attempt to fulfill his childhood dream. He had spent the best part of the past four years, as well as a small but meaningful chunk of his personal fortune, to do so. He had almost been killed a few times, too. Fortunately for Santos-Dumont, he was the son and heir of one of the richest coffee farmers in the world, so money—if not life—was without limit. He did not need to expend much valuable time away from his workshop trying to convince skeptical peers or funding agencies of the merits of his revolutionary approach to solving one of the greatest obsessions of his age.

That obsession was flight.

So it was that the flâneurs crossing the recently inaugurated Pont Alexandre III bridge happened to witness an astonishing, dead-on view of a strange flying object that seemed to be approaching the Eiffel Tower from the direction of the Bois de Boulogne. Parisians stopped wherever they stood, tracking the object's trajectory. For generations to come, these awestruck spectators would proudly brag that they were there when this son of a Brazilian coffee farmer flew his airship through the skies, like a lone soaring bird full of purpose, defeating the wind to steer its body at will. That afternoon Alberto Santos-Dumont single-handedly launched the age of controlled flight when he guided the airship he had conceived and perfected along a path circling the Eiffel Tower and then returned to the very point from which he had taken off.

As its mundane name suggests, *No. 6*, the airship that Santos-Dumont flew in his demonstration, was the sixth in a sequence of experimental dirigibles that the young Brazilian and his crew of Parisian mechanics, led by his close collaborator Albert Chapin, had built. Being the most evolved member of its class, *No. 6* included all of their innumerable aeronautical innovations.

Before Santos-Dumont, men had only left Earth's ground by ascending in spherical balloons filled with hot air, helium, or hydrogen gas. These balloons could not be steered, so their pilots had to be content to fly at the will of the wind and to use the release of ballast or gas to control the balloon's ascent and descent. As such, these aviators never controlled the balloon's trajectory or its landing place. They flew at the mercy of nature.

Santos-Dumont was clearly aware of how achieving controllable flight would reshape human life. For him, such a conquest defined a fundamental struggle, well worth the sacrifice of the inventor's own life: to liberate humanity from its "terrestrial prison" and provide the means to freely explore the far-off limits of the universe. A self-taught engineer who learned to design, improve, and improvise machines on his father's farm, he wanted to build the airships that his idol, the nineteenth-century writer Jules Verne, had dreamed would one day take man all the way to the moon.

After a succession of intermediary prototypes, Santos-Dumont invented the single-pilot dirigible, an airship that included a variety of technological advances in its shape, design, construction materials, steering technology, and instrumentation. From the beginning of his experiments with these airships, he had aimed to create a machine that could be freely driven and steered by its pilot, just as he could do when he toured France in one of the first cars to become available there. In a stroke of genius he decided to adapt petroleum car engines and add them, together with a long propeller, to his flying machines. In fact, to reach the output needed to power his airships, Santos-Dumont asked his mechanics to combine two car engines in a design that alarmed his friend Albert Chapin. Thankfully, he also had the presence of mind to bend down the engine's exhaust pipe so that the hot fumes and sparks coming off his motor would be ejected away from the dirigible's gasbag full of flammable hydrogen.

To enable the small car engines to move his dirigible, he had to find new ways to reduce the airship's weight. Being five-feet-four and very thin, his body frame was more than suitable for traveling in the smaller, lighter airships he wanted to build. He chose lightweight Japanese silk, which proved to be extremely resistant, for constructing his dirigible's gasbags, and bamboo sticks and pine wood for constructing its frame. Then, to improve the machine's aerodynamics, he departed from the classical spherical shape used in balloon gasbags and built an elongated, cigarlike gasbag that would "cut through the air."

He then struggled to find a way to control the steering of his airship at will—something no other aviator had yet experienced. After much tinkering and testing, he installed two aeronautical navigation devices that he thought would do the trick. The first of these devices was a large and movable triangular (later hexagonal) rudder, also made of Japanese silk, that was placed on top of a light frame. The frame was attached to the back of the airship's keel, or sometimes directly onto the gasbag. Santos-Dumont was able to control his dirigible's horizontal motion by manipulating a rudder. This required him to pull on a system of rope cords. Later on he adapted bicycle handlebars to serve as a steering device.

Next he turned to the issue of the airship's vertical equilibrium. As Paul Hoffman describes in his wonderful account, *Wings of Madness*, the system reflected Santos-Dumont's uncanny ability to find elegant solutions to problems that had defied a generation of aviators. According to Hoffman, the essential breakthrough came when Santos-Dumont realized that he needed to find a way to "tilt the airship, raising or lowering its nose, so that the motor would drive the balloon's ascent or descent." To do this, he would engineer "a system of movable weights by which the center of gravity of the airship could easily be shifted. The weights were merely two bags of ballast, one front and one aft, suspended from the balloon envelope." Santos-Dumont could move the dirigible's center of gravity by pulling one of the two weights into the pilot's basket. "If the front weight was drawn in, the nose of the airship would point up and if the aft weight was pulled in, the nose would point down," Hoffman explains.

As before, Santos-Dumont was willing to push the limits, including his own safety. In one prototype he went so far as to replace the traditional balloon basket with a bicycle saddle and frame, attached to a thirty-three-foot-long bamboo pole. In this configuration, people watching from a distance had the impression that Santos-Dumont was flying "like a witch sitting on a stick" that was tenuously held to the gasbag by an intricate network of cords. Not yet satisfied, he later decided to utilize piano strings instead of ropes and cords to suspend and reinforce the keel of his airship. This significantly reduced the airship's weight, and its air resistance.

At this point Santos-Dumont's dirigibles had become very complex machines, in which a multitude of ropes, handlebars, and even bicycle pedals controlled distinct piloting maneuvers. His dirigible was an unorthodox concoction of a cigar-shaped gasbag, a bamboo keel, a mesh of piano strings, two petroleum engines, a triangular silk rudder, and a system of shifting ballast bags. It's no wonder his aerial acrobatics in this engineering hodgepodge attracted attention, and that by the turn of the century he had become one of the most well known personalities in the world. Though he freely disseminated the blueprints for his airship, the collective memory of the press and Parisians served as the only record of his scientific experiments. In fact, Santos-Dumont never believed in patents. Instead, he said that his discoveries belonged to the whole of humanity.

Never shy about showing off his latest inventions to the public, Santos-Dumont regularly tested his prototype flying machines in the skies above Paris, always wearing a freshly pressed suit, a tightly knotted tie, and his favorite vapor-sculpted Panama hat. What Santos-Dumont desperately needed to silence his remaining critics was proof that this airship would fly under his control and decisively free the human body from the purely terrestrial habitat in which it had lived for millions of years.

His mission to demonstrate the possibility of man-controlled flight was suddenly boosted when the oil tycoon Henri Deutsch de la Meurthe issued a challenge to his fellow aeronautics enthusiasts. During a meeting of the Paris Aero Club in April 1900, Deutsch announced that he would pay, out of his own pocket, the sum of one hundred thousand French francs to the first airship that could take off from the Parc d'Aerostation of the Aero Club at Saint-Cloud and, without touching the ground, circumnavigate the Eiffel Tower and return to the departure point using only resources available on the airship in no more than thirty minutes. As estimated by Hoffman, to win this first official international prize in aviation, an airship would have to fly at about fourteen or fifteen miles per hour to cover the round-trip journey—a hurdle that would put even Santos-Dumont, the only credible contender for the purse, to the test.

Although many believed Deutsch's original intention was to win the prize himself, Santos-Dumont never doubted that the honor was his to lose. As a matter of fact, just before the prize was announced, the Brazilian had received permission from the Aero Club of Paris to build the infrastructure needed to fly his dirigible out of the Saint-Cloud airfield. He equipped himself with an extensive workshop for fabricating the airship's parts and a hydrogen plant for generating the balloon's gas. He also erected a large aerodrome and a balloon hangar—the first hangar ever constructed. The hangar, complete with movable doors that he had, of course, invented, gave Santos-Dumont a tremendous advantage over his competitors for the Deutsch Prize. He could now store his balloon inflated, saving the time (and money) involved in filling it with gas before each and every test flight.

For the next eighteen months, he flew all over the city, experimenting until he converged on the final design of what became *No. 6.* To get there, he had survived a couple of harrowing accidents. In the first, which occurred in July, his airship fell from the sky into the garden of Edmond de Rothschild's estate. The tree that had luckily broken the balloon's fall also proved to be a hazard. But a month later, he faced a much more serious danger. After crashing his dirigible into the side of the Trocadero Hotel, he found himself hanging for his life along some of the airship's mercifully resistant piano strings.

Not even this brush with death would deter the little Brazilian from pushing forward. Mere details seemed to separate him from a prize most of Paris already believed he had earned through sheer courage. Thus, by the time the Aero Club judging committee was summoned to observe yet another attempt at circumnavigating the Eiffel Tower on the afternoon of October 19, the only question that lingered was whether he could make the trip in thirty minutes or less. The winds of Paris were unpredictable, a nemesis that had caused him many headaches in his tests. But regardless of the conditions, his dirigible would have to perform at the outer limit of its speed, aerodynamics, and steering capabilities. Plus, he would have to survive the journey, an uncertain prospect after his experience on the wall of the Trocadero Hotel.

After an aborted attempt, Santos-Dumont's solo voyage into history began precisely at 2:42 P.M., as *No. 6* took off from the Parc Saint-Cloud and headed straight toward the Eiffel Tower. As the dirigible became visible in the city's sky, Parisians of all ranks and social position, without any ceremony, dropped whatever they were doing to secure a better vantage point for following the intrepid aviator's latest attempt. The rush of bodies and emotions produced "a wild stampede of foot-passengers, cabs, automobiles, and cyclists racing towards the Champs of Mars."

Still, according to Hoffman's reconstruction of those epic minutes, at one point the members of the French Twenty-fourth Regiment Band, who were marching on the Champs-Élysées and "serenading the visiting King of Greece and five hundred other dignitaries," heard the crowd's roaring chants of "Santos-Dumont, Santos-Dumont." Without hesitation, and taken by a synchronous collective reassurance that witnessing history more than justified the court-martial penalty for insubordination, they unceremoniously dropped their instruments and joined the running masses. About five thousand Parisians had arrived at the Trocadero Gardens by the time Santos-Dumont began to contour the "[Eiffel] Tower's lightning rod at a precariously close distance of forty feet." Aside from some slight trouble with a wind current when *No. 6* crossed the Seine, the incoming flight had been faultless. He had even set a new speed record for that part of the circuit, at eight minutes, forty-five seconds. Not having a watch to check on his timing—something his friend Louis Cartier solved a few months later by designing the very first wristwatch for him—Santos-Dumont relied on the cheers of the crowd to judge his progress and whether he was still within range to finish the trip in time. The moment he cleared the Eiffel Tower (see Fig. 9.1) and shot back toward Saint-Cloud, the huge crowds in the streets threw hats into the hair and hugged their fellow bystanders. The final dash of humanity to conquer the forbidden skies had begun.

FIGURE 9.1 Alberto Santos-Dumont and his flying machine. A photograph of the aviator (on the left) and the historic moment when he circumnavigated the Eiffel Tower on October 19, 1901. *(Reprinted with permission from the National Air and Space Museum, Smithsonian Institution Archives, Washington, D.C.)*

Despite the buoying enthusiasm below him, Santos-Dumont found the returning leg to be nothing close to trivial. Slowed dramatically by a strong headwind and three engine stalls in a row—each of which Santos-Dumont calmly fixed, midair, while still maintaining control of the dirigible direction—*No. 6* lost precious time. As he began his approach to the aeropark, he decided to make a final dive toward the ground. The ultimate seconds of the voyage were exhilarating. According to Hoffman:

> The official timekeeper clocked twenty-nine minutes and fifteen seconds. Another minute and twenty-five seconds passed while Santos-Dumont turned the airship around and brought it back to the starting point, where his workmen grabbed the guide rope and reeled him in. When his basket was low enough for his voice to be heard over the applause, he leaned over the side and yelled, "Have I won the prize?"

> Hundreds of spectators responded in unison, "Yes! Yes!" and swarmed the airship. [Santos-Dumont] was showered with flower petals that swirled like confetti. Men and women cried. The Comtesse d'Eu [the former Empress of Brazil, then living in exile in Paris] dropped to her knees, raised her hands to the heavens, and thanked God for protecting her fellow countryman. The Countess' companion, the wife of John D. Rockefeller, squealed like a school girl. A stranger presented Santos-Dumont with a small white rabbit, and another handed him a steaming cup of Brazilian coffee.

Unfortunately, the Aero Club committee could not ratify Santos-Dumont's victory at once, since—under a small but pivotal change in the prize rules enacted barely a month before—the timekeeper's clock could only be stopped when the dirigible's guide rope was grabbed by the workmen on the ground, not when the airship crossed the departure line. Santos-Dumont claimed that he had purposely overshot the line to demonstrate his disregard for what he considered a blatant attempt to make the task ever more difficult.

At long last, on November 4, the Aero Club committee awarded the Deutsch Prize to Alberto Santos-Dumont. Upon receiving it, he immediately donated half of the money to the poor of Paris, thirty thousand francs to his workmen, and twenty thousand francs to his most enthusiastic supporter, the mathematician Emmanuel Aimé. In the days that followed, Santos-Dumont acquired a hero's status, an emblematic adventurer of the new century and the anticipated new world order that would come with it. What that order would bring, nobody could tell. But at least one thing was now sure: the twentieth century would have machines that allowed people to fly at their own will and under their voluntary control. And thanks to the introduction of the telegraph and the telephone, this newsworthy event could now be transmitted to people around the world in a matter of hours.

Over the following days and weeks, articles detailing all aspects of Santos-Dumont's accomplishments filled the press reports. Almost certainly, the news reached the secluded paradise of Kitty Hawk, North Carolina, where, during the fall of 1901, two brothers from Dayton, Ohio, had returned for one more season of aeronautical experiments.

Unlike in the previous year, when their efforts were restricted to playing with a particular type of kite, during the 1901 season on the Carolina coast, Orville and Wilbur Wright spent their entire time flying a glider at the will of the wind. At no moment during that fall did the Wright brothers achieve anything remotely close to a controlled flight. That did not happen for another two years, when on December 17, 1903, the Wright brothers flew their heavier-than-air plane, *Flyer I*, from the crest of a sand dune on the same beach in North Carolina.

In my mind, there is no doubt that the Wright brothers deserve all due credit for inventing and flying the first heavier-than-air plane. Yet, if we center the question on who was the first to fly an airship that could be voluntarily controlled and navigated through the infinite skies, not as a simple and passive slave of the imposing and unpredictable moods of heaven's wind, but instead at the mercy of the voluntary motor desires of a pilot's brain, there is no doubt that the honor belongs to the Brazilian. Santos-Dumont's choice of a lighter-than-air craft to make his demonstration simply reflected his preference to follow the early aviators, including Count Ferdinand von Zeppelin, who in the summer of 1901 had preceded him in building a dirigible, but had not found a way to control it.

Needless to say, the pioneering experiments conducted by Santos-Dumont, the Wright brothers, and many others launched a technological transformation, one in which the scale of human transportation, exploration, communication, commerce, and true social and cultural integration around the globe was vastly increased. Unfortunately, so was the ability to wage war and commit horrendous crimes on a scale never before witnessed. Among the millions of victims of this transformation of airships into killing machines was none other than Santos-Dumont himself. After much emotional distress, in part from the knowledge that planes were being used by the Brazilian government to put down a civilian insurrection in São Paulo, on July 23, 1932, he took his own life.

Less than seven decades after Santos-Dumont circled the Eiffel Tower, the revolution in human reach attained a new zenith when Neil Armstrong walked on the surface of another celestial body on the very day that would have been Santos-Dumont's ninety-seventh birthday. A few years later, the late Brazilian received an even greater gift when the International Astronomical Union named a small impact crater, located in the Montes Apenninus range at the eastern edge of the vast lunar sea known as Mare Imbrium (the Sea of Rains), after him.

Santos-Dumont's flying machine illustrates in vivid and concrete terms an essential property of the human brain: namely, its ability to design, fabricate, and utilize tools to enhance its reach and the interaction of our species with the surrounding world. What neither Santos-Dumont nor his fellow aviator pioneers realized is that, around the time they were starting to venture into the forbidden skies in their flying machines, the concept of tool assimilation by the brain's "body schema" was starting to take shape in the minds of pioneering neuroscientists. In its earliest conception, posited by the British neurologists Henry Head and Gordon Holmes in 1911, a "body schema" was a creation of the human mind and, as such, this schema incorporated commonly used artificial tools. Head and Holmes had observed that patients suffering from lesions at different cortical levels of the somatosensory system experienced abnormal tactile sensations. In a presentation of their findings to the Royal College of Physicians, they proposed that the body schema helped to account for the strange perceptions reported by their patients, as well as the rich range of tactile sensations experienced by those unaffected. "We are always building up a postural model of ourselves which constantly changes," they wrote. "Every new posture or movement is recorded on the plastic schema, and the activity of the cortex brings every fresh group of sensations evoked . . . into relation with it." Head and Holmes likened this process of comparison and translation to how "on a taxi-meter the distance is presented to us already transformed into shillings and pence." In essence, the body schema was the brain's own point of view surrounding tactile information.

Head and Holmes reinforced their hypothesis by looking at cases where an extensive lesion destroyed a section of the cortex holding the original body schema. In patients who had experienced a phantom-limb sensation for many years, the loss of the schema eliminated the phantom.

If this was not shocking enough for their Victorian audience, Head and Holmes predicted that if the brain could maintain a nonexistent limb in its schema, then surely it must adopt the trappings of the body. How else could it be explained that humans had been able not only to master tools, but to feel things through them, even without the use of other senses:

> It is to the existence of these "schemata" that we owe the power of projecting our recognition of posture, movement and locality beyond the limits of our own bodies to the end of some instrument held in the hand. Without them we could not probe with a stick, nor use a spoon unless our eyes were fixed upon the plate. Anything which participates in the conscious movement of our bodies is added to the [brain] model of ourselves and becomes part of these schemata: a woman's power of localization may extend to the feather in her hat.

There is no indication in the society's notes of whether Head and Holmes's uninvited invasion of Victorian women's fashion was welcomed. But they concluded their paper on a similarly bold note, arguing that the somatosensory cortex is "the storehouse of past impressions," out of which a schematic of reality was created, often below the level of consciousness. All of our sensations, they argued, "rise into consciousness charged with a relation to something that has happened before." In fact, Head and Holmes did not simply propose that reality is born out of a body-centered cortical model of the world but directly implied that the component of this model that simulates our own physical body is our brain's ready capability to alter its spatial configuration by assimilating tools into the experience of its own flesh.

Since my first reading of their paper, I have always enjoyed the boldness with which Head and Holmes put forward their ideas. I was certainly not the first scientist to respond this way. As we saw in chapter 3, Head and Holmes's body schema reappeared in the neuroscience literature several decades later, in the shape of the "neuromatrix" theory proposed by Ronald Melzack. In the broadest terms, these theories shared a rejection of the notion of pure perception, that the neural representations of an animal's body are exclusively defined by the one-way feedforward information ascending from the periphery to the S1 cortex. Both theories also assigned to widely distributed networks of frontoparietal cortical neurons a central role in defining the familiar experience of belonging to a body. Yet neither of these challenges to the localizationist status quo made much headway for the best part of the twentieth century.

A large reason for this lack of support originated from a very fundamental shortcoming: a corresponding lack of validating experimental data. As a matter of fact, nothing in Head and Holmes's clinical cases directly supported their main hypothesis.

Moreover, as Angelo Maravita and Atsushi Iriki point out in their excellent reviews of the research on tool incorporation, Head and Holmes's original concept of a body schema was based on an unconscious integration made by the brain of a sequence of proprioceptive signals alone. By so limiting the brain's simulation to information generated within the body, they left future generations of neuroscientists with a conundrum: how could we make sense of the wealth of action potentials that seemed to arise from the large number of neurons in the frontal and parietal areas that exhibit *multimodal* receptive fields, which combine somatosensory, visual, and proprioceptive signals? One of the first steps forward was made by Ronald Melzack, even before single-neuron recordings could be taken in behaving primates, when he included tactile signals and motor activity in his neuromatrix. As both Maravita and Iriki point out, the mounting evidence for multimodal RFs, coming out of more and more sophisticated recording techniques, opened the way for testing a multimodal model of the self. In this model, "multiple fronto-parietal networks integrate information from discrete regions of the body surface and external space in a way which is functionally relevant to specific actions performed by different body parts."

At long last, sensory neurophysiologists had little choice but to realize that there was no point in studying the perceptual capabilities of an animal in a deeply anesthetized state. Because a brain's representation of the body and its relationship to the space surrounding it requires a merging of concurrent visual, somatosensory, proprioceptive, and motor signals, the new research paradigm insisted that studies had to be conducted in awake, behaving animals, preferably during experiments involving meaningful tasks. Once all the relevant information was mixed with the mnemonic traces of past experience, the subject's brain would be able to make its best prediction about the always uncertain future—just like a human brain does in real life. And only under these conditions could the neuronal circuitry of the brain generate the type of spatiotemporal patterns of activity that define the conscious experience of the self, the abstract concept usually known as "body image."

It was not until the late 1990s that the first experimental evidence was obtained to demonstrate that body image could be altered by introducing a tool into a subject's daily routine. In this innovative study conducted in 1996 by Atsushi Iriki and his collaborators at the Tokyo Medical and Dental University, Japanese macaque monkeys were trained to use a simple rake to collect a food pellet placed outside hand's reach (see Fig. 9.2). Despite not being known for using tools in their normal, outside-the-lab lives, after a couple of weeks of training, these monkeys became quite proficient in using the rake to grab and enjoy their treat. After this initial phase, Iriki and his colleagues serially recorded the activity of single neurons in the parietal cortex while the monkeys performed their newly learned trick. From the beginning, the team observed that some of the parietal cortical neurons exhibited both a somatosensory RF, located somewhere in the monkey's hand, and an equivalent visual RF, centered on the external space immediately surrounding the hand. In scientific jargon, these neurons are called "bimodal cells" because they typically respond to stimuli from two distinct sensory modalities. Since the portion of the extra-personal space adjacent to the limits of the body is usually known as peri-personal space, the bimodal neurons identified by Iriki's group primarily represented visual stimuli generated within the peri-personal space of the monkey's hand.

To their amazement, Iriki and his colleagues also observed that as the monkey moved its hand to a new location in space, the somatosensory RF of the cortical neuron remained focused on the same skin region, whereas the visual RF migrated to represent the distinct peri-personal space that now surrounded the animal's hand. The neuron's visual RF had instantaneously and properly updated the hand's position. No matter where the animal placed its hand, the cortical neurons would always keep the somatosensory and visual RFs in register. Clearly, the animal's hand position was the reference point used to define these neurons' physiological properties.

That, by itself, would have counted as a spectacular neurophysiological finding. Yet, what Iriki's group found next was even more impressive. After a monkey successfully used the rake to collect food pellets for about five minutes, the visual RF of the same bimodal cortical neurons suddenly enlarged to include the peri-personal space surrounding the entire tool, in addition to the space around the animal's hand. Plus, this dramatic enlargement of the visual field only occurred when the monkey was actively *using* the rake (in this case, to collect food pellets). If by chance the animal was simply holding the rake without actively using it, there was no change in the neuron's visual RF (see Fig. 9.2).

In these studies, Iriki also described a second class of bimodal parietal cortical neurons whose somatosensory RFs were located in the shoulder of the monkey. Before handling the tool, these neurons' visual RF included the three-dimensional, peri-personal space that the animal's unaided arm might encounter if it moved. However, following just a few minutes of playing with the rake to collect food pellets, the visual RF of the same neurons enlarged to include the entire potential, three-dimensional, peri-personal spatial range of movement of the animal's arm plus the rake. As Iriki and his colleagues rightly concluded, their data strongly suggested that the monkey's brain was assimilating the rake as an extension of the animal's arm. So precise was this assimilation that when Iriki measured the effect of rake usage on another group of bimodal cortical neurons whose tactile RF was located in the monkey's fingers, he observed no change in the corresponding visual RF. Apparently, the monkey would have to use a tool that required more specific finger movements for the visual RF to be transformed. As far as I know, no one has yet recorded from the brain of a monkey playing the violin or a piano to test this interesting hypothesis. Nevertheless, the prediction makes sense.

Iriki's group continued to turn out milestone findings in the search for the neurophysiological correlates of tool assimilation in the primate brain. For instance, in 2001 they showed that the same enlargement of visual RFs in bimodal cortical neurons could be obtained by projecting images of a rhesus monkey's hand on a video monitor while the animal manipulated objects without being able to directly see its hands, which were

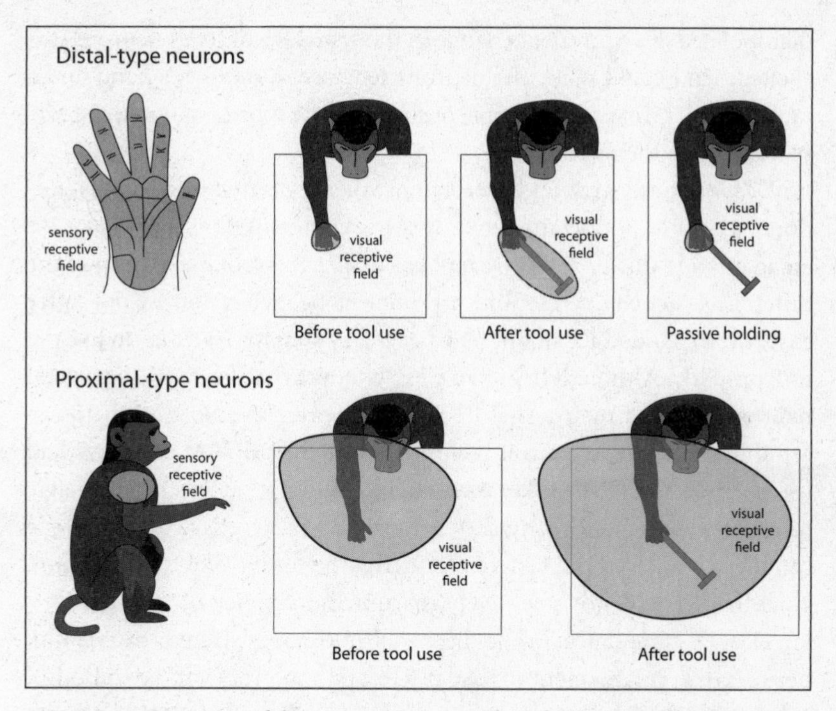

FIGURE 9.2 Summary of the experiments carried out by Dr. Atsushi Iriki and colleagues showing that the visual receptive field of a parietal cortical neuron expands when the animal employs a simple tool to perform a task. On the top shelf, a single neuron with both a tactile and visual RF centered on the hand changed its visual receptive field to include the entire tool used by the animal to retrieve some food reward. Notice that when the animal merely holds the tool, but does not utilize it to perform a task, the visual RF remains centered on the animal's hand alone. On the bottom shelf, another neuron with the tactile RF centered on the animal's shoulder and a broad visual RF shows the same expansion of the visual RF when the animal utilizes a tool in a 3-D space. Notice that the visual expansion of the RF includes the entire space in which the tool can reach. *(Adapted from A. Maravita and A. Iriki, "Tools for the Body (Schema)."* Trends in Cognitive Sciences *8, no. 2 [2004]: 79–86, with permission from Elsevier.)*

blocked from its view by an opaque barrier. Using this setup, Iriki found that the visual RF of cortical parietal neurons was centered on the video image of the animal's hand and the peri-personal space adjacent to that image. When a virtual tool was placed in the virtual hand, the visual RFs of these bimodal neurons expanded to encompass the virtual tool as expected. And when the size, position, and shape of the virtual hand were manipulated, an equivalent change in the visual RF of those neurons was elicited. Iriki could make the neurons represent a monkey's hand and a tool held by it in any way, shape, or form in which he configured the virtual image on the monitor.

Despite these striking observations, there was one important question that could not be answered: was the enlargement of the visual RF an adaptation due to *repetitive* tool use or *effective* tool use? This was the difference between a rote and a proficient behavior, and in the latter case, the expanded RF might be a necessary step for learning to use the tool properly. Although Iriki had clearly shown that the cortical parietal neurons changed their visual RFs to incorporate the tool and the space around it, his single neuron recordings only began after monkeys had learned how to use the rake. As such, he could not say when the change took place, whether during the monkeys' training phase or just as the monkeys mastered the task of retrieving the food pellets. In addition, since Iriki could not selectively disrupt the activity of this particular class of bimodal neurons, he could not demonstrate a causal link between the enlargement of visual RFs and the proficient operation of an artificial device. Consequently, one could still argue that the enlargement of the visual RFs was a mere consequence of any tool use, no matter how skilled it was.

To be fair to Atsushi Iriki and his collaborators, the task of establishing a causal link between an observed pattern of brain activity and the production of a particular behavior is one of the most difficult challenges a neurophysiologist faces. In this specific case, however, the data on Aurora's BMI, which we published a couple of years after Iriki's paper appeared, shed some light on this unanswered question.

As we saw in chapter 7, the velocity and direction-tuning profiles of individual cortical neurons reveal two interesting classes of neurons that could be found in every one of Aurora's cortical areas. One of these neuronal groups displayed velocity and direction-tuning properties that were somewhat similar when Aurora used either her own biological arm or the external robotic arm to play the video game. The second class of neurons, though, only exhibited clear velocity and direction tuning when

Aurora was controlling the robot arm movements using her thinking alone. This enhanced neuronal firing, which showed clear and sharp velocity and direction tuning, occurred a few hundred milliseconds before the onset of the robot arm's movements, which is why the BMI allowed Aurora to intercept the target with the cursor and win gallons of delicious fruit juice. Indeed, without these two classes of cortical neurons, our BMI simply could not work; the remaining neurons recorded from Aurora's cortex tended to shut down when she stopped making movements with her arm. Interestingly, the moment we turned off the BMI and Aurora returned to using her arm to play the game, the cortical neurons that fired prior to the movements of the robot stopped firing altogether. Unlike in Iriki's experiments, our recordings had been initiated from the very start of Aurora's training with the BMI.

As Aurora transitioned from moving her own arm to controlling the robot arm by thinking alone, the ninety-six cortical neurons we were simultaneously recording displayed a three- to sixfold increase in the covariance of their firing, measured in sequential one-hundred-millisecond bins. This meant that the direct operation of the robot arm by this random sample of neurons was correlated with a distinct, broad temporal pattern of similar firing among these neurons. To make things even more intriguing, this effect was not restricted to neighboring neurons. Instead, it extended to groups of cortical cells that were far apart, as if by reducing the variability of their individual firing times, these neurons could create closely related, though spatially dispersed, circuits that shared a similar task. Yet again, the moment the BMI was turned off and Aurora returned to using her own arm while playing the video game, the neuronal firing covariance structure broke off, too, and the firing across the recorded sample returned to the original, temporally dispersed pattern. This use of time to "glue" space appeared to be a trick of the brain to allow widely dispersed populations of neurons to interact, ever so briefly, during the generation of a fundamental exploratory behavior.

Although some neurophysiologists could argue that the combined findings from Iriki's and my own lab do not constitute definitive proof that observed changes in cortical neuronal firing patterns and RFs determine the ability of monkeys to become proficient tool users, our research certainly pointed in that direction. We gained some extra support from a series of studies conducted with humans. For example, Lucia Riggio and her colleagues at the University of Padua have demonstrated that humans can distinguish a tactile stimulus equally well, regardless of whether it is delivered at the finger's skin surface or at the end of a long tool. In another study, conducted at the University of Milan, Angelo Maravita and his team observed that a visual distractor, such as a flashing light, interfered with receiving another sensory signal, in this case a tactile stimulation, regardless of whether the subjects were using their hands or long tools to report what they felt. More recently, a group led by Lucilla Cardinali at INSERM (the National Institute for Health and Medical Research) and the Université Claude Bernard in Lyon, France, reported that using a mechanical grabber to perform a simple task significantly alters the kinematic properties of subsequent free-hand movements. The study revealed that, after use of this tool, people misreported the length of their own biological arms by indicating that touches simultaneously applied to their elbow and middle finger were further separated than in the period prior to tool utilization—implying that using the tool gave the subjects a feeling that their arms had been physically extended. Can you imagine what arm length Roger Federer would report following an hours-long battle employing his customized racket against his archrival, Rafael Nadal?

Further support for the thesis that tools readily become a part of our internal, brain-based self-representation comes from a series of clinical cases involving patients who have suffered extensive cortical lesions. For instance, Anna Berti of the University of Turin, and Francesca Frassinetti, of the University of Bologna, both in Italy, report that, following a massive stroke, one patient had developed a huge lesion in the right cortical hemisphere, which limited the blood flow to a large area of the right parietal lobe. The patient developed a hemineglect syndrome, ignoring the leftmost side of both her body and the world—down to the leftmost letters in a word and the leftmost words in a sentence she was asked to read. Upon a closer examination, conducted a month after the cortical event, the neurologists discovered that the patient's hemineglect had evolved. When an object was placed about twenty inches away from the left side of her body, the patient could not perceive its presence; however, if the same object was placed farther away, about forty inches, the patient could suddenly sense it. The hemineglect had been restricted to a spatial terrain quite near to her body.

To further characterize the hemineglect syndrome, the research team next tested the patient on a line bisection task. In this task, straight lines were drawn on a piece of paper mounted on a board and placed in different spatial positions in front of the patient. Then the patient was asked to use her right index finger to point to the middle of a straight line positioned in her neglected space. Typically, patients suffering from left spatial neglect displayed a rightward bias when trying to bisect a straight line positioned on their left side. In their examination, however, Berti and Frassinetti introduced two new ways in which the patient could bisect the line, in addition to using her index finger: for nearby lines she could now use a laser pointer, and for lines placed farther from her body she could use either a laser pointer or a forty-inch-long stick. As predicted, when the thin black line was placed on the patient's near-left extra-personal space, she misjudged the middle point of the line, whether using her fingertip or a laser pointer, by placing it much farther to the right. Yet, when the line was placed in the far-left extra-personal space, the patient could bisect the line flawlessly while using the laser pointer (as she did when she used her own right finger), but committed the same rightward errors when she used the forty-inch-long stick. Berti and Frassinetti proposed that these latter errors occurred because, through the use and incorporation of the long stick, the patient's brain had somehow converted the far-left space into near-left space, a region in which the patient's spatial neglect fully manifested itself.

All together, the neurophysiological, psychophysical, and clinical findings I have thus far described constitute a small sample of the vast scientific literature that upholds, with a great deal of certainty, Head and Holmes's original hypothesis of tool assimilation into the brain's model of the body. Although many animals, from insects to mammals, exhibit some capability to utilize natural tools or artifacts to their advantage—a phenomenon Richard Dawkins calls the "extended phenotype"—several million years of evolution endowed the human central nervous with the ability to outperform all other species. By harnessing a precious combination of cortical algorithms with the dexterous multimodal sensorimotor ability to grasp, reach, and manipulate a created tool and dynamically represent extra-personal space around an ever-moving body, the human brain has produced the most adaptive and complex model known to us: one's sense of self.

The overall impact of this relativistic, brain-derived model is tremendous. Out of this explosive catalytic mix of mental creativity, motor dexterity, and unending assimilation of extra-personal space and artifacts, for the past few million years the human species has gradually climbed a very unusual evolutionary path. Far from merely allowing us to develop new technologies that extend our individual and collective reach, enlarge our habitat, and enhance our means of producing food, surviving diseases, and enduring natural disasters, this exuberant brain-based simulation assures that all of our future technological developments, whatever they may be, will be continuously and actively assimilated into the sense of self of each generation to come. That may sound outlandish to some, but after spending the past two decades reviewing the accumulated findings of a century of neurophysiological research, I found this to be the only parsimonious way to describe how brains generate our ever-shifting sense of being. In fact, I would go even further. Because of its unsurpassable and pragmatically effortless powers to employ and disguise complex artifacts as extensions of our human flesh, the human brain possesses the only biological algorithm capable of wrestling away from our genes a significant chunk of responsibility for defining the future of human evolution.

It will take a multitude of experiments and many years of debate to test the validity of my theoretical assertion, but the weight of the evidence collected thus far points like a beacon in this direction. For instance, we have already seen that experimental evidence supports the notion that one's sense of self is not limited to the outermost layer of the epithelium. Instead, it likely comprises the clothing, wristwatch, rings, socks, tie, gloves, shoes, hearing aids, dental fillings, prosthetic limbs, pacemaker, glasses, contact lenses, fake nails, wig, dentures, artificial eyes, necklaces, earrings, bracelets, body piercings, silicone implants, and every other add-on applied to or within the body. Furthermore, one's sense of self likely encompasses all the tools we commonly, and not so commonly, utilize, directly or remotely, provided the movements of these tools are somewhat correlated with the movements of one's own body parts. For most people, one's sense of self surreptitiously expands, over the course of a lifetime, to include technological tools with which we are actively engaged, such as a car, bicycle, motorcycle, or walking stick; a pencil or a pen; a fork, knife, spoon, whisk, or spatula; a tennis racket, golf club, basketball, baseball glove, or football; a screwdriver or hammer; a joystick or computer mouse; and even a TV remote control or a BlackBerry, no matter how weird that may sound.

For a subset of people with specialized proficiency, the self can enlarge to include a musical instrument, say a violin, piano, flute, or guitar; a medical implement such as a scalpel or a microelectrode; or a transportation device along the lines of a sailboat, backhoe loader, lunar capsule, or plane. Which is why Alberto Santos-Dumont's adventures with his balloons, dirigibles, and planes can tell us a great deal about tool assimilation.

According to Santos-Dumont's own written and oral reports, by continuously pulling, with his own hands, the many ropes that controlled the components of his dirigibles, and immediately experiencing the airship's steering responses as he traveled through the Parisian sky, he began to feel as if the airship's movements were his own. That was quite a departure from the sensation he had experienced in his previous flights as a passenger in the basket of a traditional balloon that moved at the mercy of the wind. Santos-Dumont's sense of self became so entangled with his flying machines that, later on, when he was working in 1908 on the blueprints for the *Demoiselle*, the most reliable single-pilot heavier-than-air plane of its time, he made sure to create ways in which different parts of his body could be directly attached to the specific controls of his beloved new airship. Thus, by simply moving his body back and forth or laterally, he could steer his plane. Knowing what we know today, one can speculate that Santos-Dumont began to experience the

FIGURE 9.3 The left panel shows Pelé in one of his characteristic shooting maneuvers in the 1960s. On the right, a drawing represents how the theory proposed in this book predicts what Pelé's sensorimotor cortex would look like. According to this view, the soccer ball would be incorporated into Pelé's foot representation in the cortex. *(From Swedish press image in the public domain.)*

first closed-control feedforward/feedback loop among millions of interconnected human neurons and a flying machine, and that in the process his brain incorporated the entire plane into its image of his petite body. The exquisite level of tactile sensitivity that Santos-Dumont experienced most likely rivaled the mastery of the Formula 1 racers Ayrton Senna and Nelson Piquet, who said they were able to detect minute changes in the asphalt surface of a racetrack while driving their cars at more than 150 miles per hour. A similar process may explain why Pelé rarely looked directly at a moving soccer ball before initiating a dribble, pass, or commanding shot into the goal. The brain of the greatest of all soccer players had long assumed that the football was nothing but an extension of his moving feet (see Fig. 9.3).

These master tool wielders incorporated their tools into their brains' body image. Thus, as Santos-Dumont flew *No. 6* or his "petite *Demoiselle*," and as Pelé took possession of the ball in each of the 1,363 professional games he played, each man's brain newly adopted the respective tools into his body schema, adjusting his sense of self and the related sensory receptive fields in real time. This process encompasses yet another neurophysiological principle of the relativistic brain, the *plasticity principle*.

THE PLASTICITY PRINCIPLE

The representation of the world created by populations of cortical neurons is not fixed, but remains in flux, throughout our lives, continuously adapting itself to new learned experiences, new models of the self, new simulations of the outside world, and newly assimilated tools.

The plasticity principle encompasses all of the mechanisms of cortical reorganization that underlie the ability of animals to learn new tasks, including the incorporation of artificial tools as expansions of the internal models of one's self. As such, it constitutes one of the main reasons brain-machine interfaces actually work—the brain does not discriminate between a tool that is, for example, sensed in a biological hand and a virtual one.

At this point, I have to admit that my focus so far on the tools incorporated by the brain does not completely satisfy me. In my sincere opinion, one's sense of self can reach far beyond these limits. Although the experimental evidence remains scant, I firmly believe that, in their perfectionist drive to achieve the ultimate simulation of the self, our brains also incorporate, as a true part of each of us, the bodies of the other living things that surround us in our daily existence. The refined neural simulation routine I am proposing may be better understood if we call its final product by its more colloquial and well-publicized name: love.

Think about how love, and its more intense incarnation, passion, arise in us. There's love at first sight, the murmur of sweet nothings, a mother's tender embrace. Each of these involves classical sensory channels (visual, auditory, and tactile). There's also the cascade of hormones that come with the ever-sensitive chemical senses, olfaction and taste, the examples for which I will leave to your extraordinary human imagination. The brain receives a continuous flow of multimodal signals in each of these situations and strives to incorporate that flow into its existing model of reality and its sense of self, built on its prior experiences, just as it would when faced with the feedforward and feedback information of a video game joystick or a BMI. In this approach to defining the relativistic brain, one's sense of self would have to include parents, spouses, and children, and to a lesser degree, relatives, friends, and perhaps minor social acquaintances. Even our pets could be part of this list.

There are some indirect hints, obtained by the investigation of social behaviors in our species as well as other mammals, that lend some initial support for this far-fetched idea that the brain assimilates other living bodies into its internal image. Take, for example, the behavior of the North American prairie vole. When a young adult of this species encounters a mate it really likes, its brain tends to release a great deal of dopamine, a molecule that mediates a strong, pleasant, and rewarding sensation. After this initial passionate meeting, a typical prairie vole establishes a very strong social bond with its partner, leading to a relationship that usually lasts a whole life. Thus, despite the fact that both female and male prairie voles continue to engage in fortuitous sexual escapades with a variety of other individuals, they maintain a much stronger bond with a particular companion with which they live. Further studies have shown that when prairie vole couples are in direct social contact, they produce high levels of the hormone oxytocin, the same chemical released when women breast-feed their infants. Once released, oxytocin can bind to specific receptors located in the limbic areas of the brain and induce dopamine release. Consequently, a well-acquainted prairie vole couple is likely to experience a very pleasant, long-lasting feeling of reward, something they do not necessarily experience in their briefer encounters with occasional sex partners. Interestingly, blocking the oxytocin receptors in a female vole who has just delivered a litter makes it "socially blind" to its new offspring. It follows, therefore, that if oxytocin reception was blocked in a prairie vole couple, the animals would break their bond and fill their lives with a never-ending series of one-night stands.

While that's all well and good for prairie voles, brain imaging studies carried out in young human couples who report being in the early stages of a passionate romantic relationship have revealed a similarly intense activation of dopamine-rich areas of the brain. Moreover, it is now known that oxytocin is released when people embrace their loved ones, such as a spouse or child, when couples have sex, and even when one of us meets up with a close and beloved friend. Oxytocin release may also contribute to the pleasant feeling people experience when they caress a pet or enjoy a massage. These and other observations have raised the possibility that oxytocin, among other hormones, plays a key role in intermediating social human bonding through a sequence of positively reinforcing, hedonic responses. These responses would be initially triggered by behaviors that elicit physical body contact—things like hand holding, kissing, hugging, and sex—or social encounters with an object of desire. Then, the hormones and chemicals released by the brain would create an intensely pleasant sensation, eventually establishing a long-lasting bond. This bond would be nurtured by the brain's simulated reality, until it became integrated as part of the very model that defines one's sense of self.

In my view, this perceptual-chemical transduction mechanism for body assimilation and plasticity could define a causal chain of events through which the human brain expands its neural model. Surprising as it may sound, this would imply that our sense of self also encompasses a vivid representation of the social network of individuals with whom we share our lives, a true amalgam of bodies that is actively and dynamically maintained in neuronal space by a crowd of touches, embraces, kisses, and caresses distributed to those we love. This might even explain in neurophysiological terms why it is so painful to cope with the end of a romantic relationship or the death of a loved one. Basically, I propose that such a terrible, all-consuming pain emerges because, for our ever-meticulous modeling brains, such a loss truly represents an irrevocable abdication of an integral part of the self.

But is the assimilation of other living beings the final limit to which one can stretch the sense of self? Precarious as this may sound, I believe the answer is a resounding no. The emergence of brain-machine interfaces, in combination with new technologies for remote operation, or tele-operation, of the most varied sort of mechanical, electronic, and virtual tools located at considerable distances, or even in different spatial scales, from the physical presence of their biological operators, all but guarantees that the unique ability of the human brain to assimilate the tools it creates has the potential to expand the limits of the self to domains never before visited by any representative of our species.

What I have in mind goes much further than Aurora's ready assimilation of a robot arm, located a few yards away, or Idoya's assimilation of a pair of robotic legs, located on the other side of the planet. For instance, could the self be extended so that it would be possible for it to experience the unfamiliar surface of another planet through a mechanical device that has been sent there years, decades, or centuries earlier? It is at least theoretically conceivable that such an inconceivable frontier may be crossed in a generation or two. And when the time comes, it is plausible—in fact, it is almost certain—that our grandchildren will not comprehend why the earthbound generations that preceded them thought it was so stunning to wander the red dunes of Mars, and feel the cold, flaky sand being pushed underfoot, from the comfort of a chair in their own terrestrial living rooms.

10

SHAPING AND SHARING MINDS

"Has anyone tried this before and survived peer review to tell the story?" Through an irritating muffling caused by the international phone call's background noise, I could detect some polite hesitation in certifying my latest unconventional idea. Having stayed awake the whole night sketching out an experimental setup I had been ruminating over for months, I now faced the real struggle: getting someone else to believe my strategy would work.

It wasn't going to be an easy sell.

"Do you really think a brain-to-brain interface can be built? Connecting two living brains? Hard to believe!" Although my interlocutor seemed mildly alarmed, he was not quite ready to move to safer ground, like, for instance, our favorite topic: the status of the Palmeiras soccer team. That was my first good sign.

As I had done many times in the past thirty years, I was testing the waters with my childhood friend Luiz Antonio Baccalá. This time, however, I had not chosen this protocol purely out of safe habit. An accomplished and intellectually gifted electrical engineer with a PhD from the University of Pennsylvania, Baccalá can dissect the most complex scientific theory with a crystalline sharpness.

"I will fax you some drawings of what I have in mind," I replied. Now my friend knew I was not joking. Drawings had always been a trial for me. But I had realized they would be the fastest way to convey my ideas for a brain-to-brain interface.

"No problem. Send me the drawings. I will let you know what I think as soon as I have some time available. I may have time this weekend. Or maybe not. I will see what I can do."

As the call ended, without a single mention of Palmeiras's customary mistreatment of its loyal fans, I grasped that Baccalá was intrigued.

Other people had toyed with the idea of connecting the minds of two people. For instance, in his 1994 book *The Quark and the Jaguar*, the Nobel Prize–winning physicist Murray Gell-Mann wrote:

> Some day, for better or for worse . . . a human being could be wired directly to an advanced computer (not through spoken language or an interface like a console), and by means of that computer to one or more other human beings. Thoughts and feelings would be completely shared, with none of the selectivity or deception that language permits. . . . I am not sure that I would recommend such a procedure at all (although if everything went well it might alleviate some of our most intractable human problems). But it would certainly create a new form of complex adaptive system, a true composite of many human beings.

Having convinced myself that Gell-Mann's fear was unjustified, and that in the future such a technology could be extremely beneficial to humanity (see chapter 13), I devoted a great deal of effort to creating a legitimate strategy for testing a true brain-to-brain interface. Now I just awaited Baccalá's seal of approval to unleash the project.

The following Monday, I spotted an e-mail message, sent early that morning, from Dr. Luiz Antonio Baccalá, professor at the Escola Politecnica, University of São Paulo: "Call me at once." Besides this urgent plea, nothing else was written.

As usual, he did not answer his cell phone. After a few tries, I got lucky and reached him at his office. Sounding very irritated, he started by explaining that he did not have much time to talk; he was busy grading student exams. But he needed to tell me something important. He then fell into a deep silence. The next few moments stretched unbearably.

"What is it?" I begged.

"It is crazy, in the good sense of the word: highly unpredictable, disruptive, but extremely attractive. If it works, nothing will be the same in your field. If it does not work, there's nothing to lose, except perhaps your hard-earned reputation. But that is meaningless compared to what will happen if it does work."

Coming from Baccalá, such an unconditional vote of support meant a lot. He had carefully examined the drawings, analyzed the technical issues. For him, the logic of the experiments was sound. And that was all I needed to hear.

Contemporary neuroengineering is quickly approaching the ability to link two and even many brains together. As we have seen, the successful operation of a BMI dedicated to controlling the movements of a machine requires the implementation of two concurrent components: one that samples brain activity, extracts voluntary motor information from it, and redirects the resulting command signals to an artificial device (the efferent component), and another that provides feedback information describing the performance of the actuator back to the subject's brain (the afferent component). Most of my descriptions of BMI experiments so far have focused on the first component and the ways in which brain-derived signals could be employed to extend the reach of the human brain through the assimilation of artificial tools. In most of these examples, direct or remote visual feedback from the machine's movements was used in the second component of the BMIs. Although vision plays a fundamental role in the brain's natural process of tool incorporation, BMIs that rely on other sensory modalities have been built and operated. In fact, the most successful neuroprosthetic device ever built is the cochlear implant, which has already allowed tens of thousands of deaf patients worldwide to regain a significant functional level of hearing. The implant employs electrical stimulation of the remaining fibers of the auditory nerve to elicit its clinical effects.

If anything, it is fair to say that visual feedback became the preferred choice for feedback in BMIs because it is easy to implement in a laboratory setting. Nonhuman primates handle visual feedback very efficiently and exhibit no difficulty interacting with video screens. But there is no reason why other sensory modalities should not be equally utilized. In fact, in recent years, Nathan Fitzsimmons and another graduate student at Duke, Joseph (Joey) O'Doherty, have shown that tactile stimulation of a monkey's skin can readily substitute for visual feedback as the afferent component of an upper-limb BMI like the one operated by Aurora. For instance, in the presence of ambiguous visual information, monkeys easily learn how to use tactile cues to decide in which direction to move the robotic arm being controlled directly by their thinking.

Still, such a method of delivering feedback information relies on the highly specialized sensory apparatuses of the body. It's hard to put forward a credible claim that BMIs thoroughly liberate the brain from the body when that's the case. To truly push beyond these limits, one would have to find a way to implement the BMI's afferent feedback component without any intermediation by the body's peripheral sense organs.

It so happens that a few key modifications of electrical brain stimulation—the technique used by Eduard Hitzig and Gustav Fritsch when they discovered the motor cortex in 1870, and one of the most common experimental approaches used by neurophysiologists in the past century—offer a very convenient starting point for solving this dilemma. In our first attempt to create a brain-machine-brain interface (BMBI) for motor control, we decided to adapt this method to communicate directly with the brains of our monkeys and then investigate whether they could learn to decode simple instructional or sensory feedback messages delivered directly to the cortex. Though we were using a typical BMI at this stage, we adopted the term *brain-machine-brain interface* from a 1969 study that first outlined a device that would allow bilateral communication between a subject's brain and an artificial actuator without any interference from the subject's body. The original implementation involved an automatic interaction between two subcortical brain areas, mediated by an analog computer. Our BMBI required the subject to voluntarily control the device through a well-defined motor task.

More than a hundred years of electrical brain stimulation experiments offered not only inspiration but also a vast repertoire of practical tips in applying electrical brain stimulation to our problem. After all, most leading neuroscientists, including Sir Charles Sherrington, Sir

Edgar Adrian, and Wilder Penfield, had tinkered, in one way or another, with electrical stimulation to probe different parts of the central and peripheral nervous system. Yet none of these giants took the technique as far as the largely forgotten Spanish neurophysiologist José Manuel Rodriguez Delgado, who deserves most of the credit for launching the modern era of using chronic brain implants in freely behaving animals and humans from his lab at Yale University.

In one of his favorite experiments, carried out in 1969, Delgado demonstrated the automatic operation of the first bidirectional BMBI using Paddy, a female rhesus monkey, and a tiny device he had invented, the "stimoceiver," which allowed wireless radio transmission of electrical signals between the brain of a freely behaving subject and a machine. Because of its size, multiple stimoceivers could be implanted at the same time, so distinct brain regions could be simultaneously recorded and stimulated. In his experiments, Delgado relied on chronically implanted EEG-recording electrodes to sample the electrical activity produced by neurons located in the almond-sized deep brain structure named the amygdala that appears to be involved in regulating emotions. The stimoceivers relayed the amygdala's raw brain signals to an analog computer, located in a room adjacent to Delgado's lab. By programming this computer to detect a particular pattern of rhythmic brain activity, the so-called amygdala spindles, that resulted from the activation of a coherent ensemble of amygdala neurons, Delgado set a clear criterion for triggering the feedback component of his BMBI: every time an amygdala spindle was detected, the computer sent a radio signal to the stimoceiver instructing it to trigger an electrical stimulation of a separate brain area, a portion of the monkey's reticular formation, which he had previously identified would inflict a negative reinforcement mechanism.

This cunning arrangement allowed Delgado to set the BMBI in motion and simply observe how the interactions between these two subcortical brain areas, mediated by a machine, played out. To his astonishment, just a few hours after initiating the BMBI, Delgado observed a 50 percent reduction in amygdala spindle activity. Over the next six days, Paddy spent two hours each day interacting with the BMBI. By the end of the period, her amygdala spindling had been reduced to an incredible 1 percent of its normal levels. At this point Paddy became quieter, withdrawn, and less motivated to participate in further behavioral testing, and so Delgado discontinued the experiment. Within a few days her amygdala spindles, as well as her cheerful demeanor, bounced back to normal levels. This led Delgado to predict that physicians in the not-so-distant future would directly link the human brain with computers to treat neurological disorders.

Sadly, it wasn't long before Delgado and his research met with scientific ostracism. According to a 2005 *Scientific American* article by John Horgan, Delgado evinced strong, almost visceral antagonism from his peers and the lay public alike. Not particularly known for his subtlety, he certainly did not shore up his reputation by choosing *Physical Control of the Mind: Toward a Psychocivilized Society* as the title of the book he published in 1969 to summarize his experimental findings. But it was his particular vision for the future of brain implants—using them to modulate physiological and pathological behaviors in both animals and humans—that stoked fear that neuroscientists were gaining the knowledge and technology necessary to craft an effective method for mind control. After all, he was working at the peak of cold war paranoia, and any conspiracy theory, particularly one involving a scientist messing with someone's mind, sounded terrifyingly plausible. Had the scaremongers read Delgado's book, however, they would have learned that his experiments had daringly explored using intracranial electrical stimulation to probe cortical and subcortical circuits, primarily for the purposes of acquiring basic knowledge of the brain and, perhaps eventually, developing clinical therapies for severely ill patients. There is no doubt, though, as Horgan notes, that Delgado was fascinated with the possibility of finding a way to communicate directly with the human brain.

I still recall the day, in the fall of 1994, when I first pulled *Physical Control of the Mind* off a library shelf that seemed to be visited exclusively by spiders and the occasional homeless termite. Having just begun my career as an assistant professor at Duke, I had decided it was about time that I read some of the classic books of neuroscience. My curiosity had been stirred when I heard that Delgado had staged an experiment on the inhibitory behavioral effects of electrical brain stimulation in a bullfighting arena, of all places (see Fig. 10.1).

This fantastic setting is brought to life through a quick sequence of black-and-white photographs taken at a ranch in Cordoba, Spain. A lean, mean bull, whose ancestors had been carefully bred for many generations to enhance a single personality trait—a ferocious dislike of men holding red capes—plays out its scripted role, charging at full speed at the neurophysiologist who, at first glance, seems to be armed only with the typical red dress cape worn by the archetypal matador, the gold-swathed torero who, in Bizet's immortal opera, manages to steal Carmen from Don José, every time.

Initially, the enormous bull stands at the edge of the arena, aiming its lethally sharp horns at Delgado, who stares at his experimental subject attentively while valiantly holding his red cape with his right hand. In his left hand Delgado holds an object that had likely never before been seen in a *corrida de toros*: a radio-like device sporting a long antenna. Along the arena's wooden ring sits a mysterious helper, who displays no tense sign of concern or distress about the events that are about to unfold. The bull next launches its seemingly irreversible charge, its horns pointed directly at the amateur torero's torso. Nothing can prepare one for the relief and surprise when the next snapshot reveals the bull sliding to a halt, just a couple of yards away from Delgado who, a mere second earlier, has wisely dropped his useless cape and, without releasing his eyes from the hurtling bull, shifted his entire motor will (and his prayers) to the task of pressing a button on his strange device. The now tamed bull turns away from Delgado, who can be seen waving the shamed animal away with his right hand. It's then that one appreciates that the helper has hardly moved a muscle throughout the entire frame-by-frame account, betraying that this was, to him, a rather pedestrian exhibition of a well-rehearsed trick. If it had not been, as Dr. César Timo-Iaria liked to say, the historical record of that amazing experimental encounter would have exposed a trembling graduate student, rather than Delgado, as the courageous torero in custody of the radio-like equipment.

No matter who was controlling the device, Delgado had found a way to stop a charging bull, just moments before the animal was set to gore him, apparently by pushing a single button. That, by itself, looked incredible. In fact, what Delgado demonstrated that day, in a rather extravagant way, was that by electrically stimulating particular regions in the bull's brain, including the striatum of the basal ganglia, a major

FIGURE 10.1
The neuroscientist in the bull ring. This sequence of photographs illustrates the classical experiment carried out by Dr. José Delgado in which a charging bull was made to stop its charge using deep electrical brain stimulation. *(Photographs used with permission of Dr. José Delgado.)*

conduit for motor information, he was able to induce a state of "motor behavior arrest" in the animal. Being an ingenious technologist, he had used a radio-wave frequency to activate a stimoceiver he had previously implanted in the bull's brain.

Delgado's unorthodox experiments with electrical brain stimulation did not end in the bullfighting arena. He was the first to study how the use of electrical stimulation to contain the aggressive behavior of the dominant "alpha" monkey of a colony affected the status of that monkey and other members of its social group. In monkey colonies, a single alpha male imposes its will on the lower-ranking "delta" members by displaying a range of threatening behaviors, such as staring directly at other animals, baring its teeth, emitting a vocal warning, and changing its physical posture to indicate a potential attack. Even in captivity, this behavioral arsenal ensures that the alpha monkey maintains key privileges in the colony, including much of the cage space, a choice of the females for mating, and first access to the food provided by caretakers. As Mel Brooks might say, it's no fun being a delta monkey.

To start, Delgado implanted his usual stimulating electrode gear in Ali, the dominant alpha monkey. This allowed Delgado to remotely stimulate the caudate nucleus—a brain region that is associated with motor control—of Ali's basal ganglia as the monkey interacted with delta members of his colony. During one hour each day, Ali's brain was stimulated for five seconds, once a minute, and during this hour of intermittent stimulation, Ali's aggressiveness diminished dramatically. As Ali's new demeanor was gradually recognized by the rest of the colony members, the lower-ranking monkeys started to assert themselves and moved to strip Ali of his territorial and other privileges. While Ali's brain was being stimulated, the delta monkeys spread themselves around the cage, even crowding around Ali, who did not seem to care about what, at other times, would have been a severe affront to the colony's strict hierarchical order, an act worthy of punishment.

The relaxed gang of delta monkeys did not get to enjoy their party for long, however. About ten minutes after Delgado terminated the stimulation of Ali's brain, the old order was readily reestablished. Back to his usual aggressive self, Ali reconquered his territory and all the sweet perquisites that came with being an alpha primate. In a subsequent series of experiments, Delgado decided to measure what would happen to the social structure of the monkey colony if lower-ranking animals had access to a lever that, when pressed, triggered the electrical stimulation of Ali's caudate nucleus. At first, a few of the delta monkeys pressed the lever tentatively. After a while, though, a female named Elsa discovered that every time Ali threatened her, she could press the lever to get him off her case. Elsa's own behavior also evolved; she soon began to stare directly at Ali when she pressed the lever. Although she did not become the new alpha monkey of the colony, Elsa clearly gained control over Ali's aggressiveness and kept his attacks to a minimum.

During his career, Delgado employed the same general method to electrically stimulate a large variety of cortical and subcortical brain structures related to key brain circuits, such as the motor and the limbic systems, in animals and up to twenty-five human subjects suffering from severe psychiatric and neurological disorders. He could induce and block a range of behaviors: complex motor actions, perceptual sensations, emotions such as aggression or affinity toward others, euphoria, tameness, and lust. Through these studies, he swiftly realized the many pitfalls involved when employing electrical brain stimulation to produce a particular behavior. As Horgan notes in his profile, Delgado dropped most of his research on humans when he observed that the effects of electrical brain stimulation varied greatly from case to case and even in the same patient, from moment to moment.

It is easy to imagine how this kind of research eventually got Delgado into very hot water. Indeed, as Horgan suggests, these experiments look frighteningly reminiscent of the worst possible scenarios peddled by science fiction writers. However, as Delgado revealed in an interview with Horgan, his main intention in pursuing electrical brain stimulation in humans was primarily motivated by the fact that, at the time, schizophrenic patients who exhibited bouts of aggressive behavior were commonly subjected to a gruesome surgical procedure—the prefrontal lobotomy—that involved disconnecting from the rest of the brain, removing, or destroying most of the patient's prefrontal cortex. Tragically, it took many years for neuroscientists to uncover that the lobotomy reduced the patient to a mental state characterized by emotional indifference, lethargy, a profound indifference to pain and other feelings, and a marked lack of initiative and drive. Delgado was horrified by the procedure. Such good intentions alone, however, did not appease his detractors in the scientific community. Nor did they calm members of the general public, who started to question the aims of his work, on many fronts and on multiple grounds.

By 1974, just five years after his controversial choice for a book title had obfuscated most of his technological and scientific discoveries, Delgado left the United States to accept a position created specially for him at the Universidad Autónoma de Madrid, in Spain. There, he continued his work, isolated from much of the mainstream neuroscience community, focusing primarily on noninvasive methods to stimulate the brain. His experiments, particularly those involving the generation and blockade of motor behaviors, indirectly paved the way for the introduction, a generation later, of deep brain stimulation (DBS) in the treatment of Parkinson's and other neurological diseases. Yet, for the next two decades, Delgado's name and legacy slowly vanished from the neuroscience literature.

Curiously, nothing Delgado did or published ever suggested that he would become capable of controlling someone's voluntary will, let alone someone's mind. It is interesting to note, though, that the usual suspects who routinely denounce the supposedly innumerable creative ways in which scientists are contributing to the extinction of human nature remain utterly untroubled by the repetitive demonstrations that much more efficient forms of brainwashing and mind control abound in modern society. Contrary to these imaginative plots, none of these entail any last-generation brain chip developed in the laboratory of a neuroscientist, even one as eccentric as José Delgado.

Despite Delgado's troubles, electrical brain stimulation continued to be widely employed by neurophysiologists as a key method to stimulate brain tissue, neural pathways, and peripheral nerves. However, for a long time few neuroscientists dared to come close to the type of experiments Delgado performed in the 1960s. That placid scenario took a drastic turn, however, when my former adviser, John Chapin, and his students stunned the neuroscience community with their extensive experiments with "roborat" (see Fig. 10.2).

Although I had known about their work from the time of their initial conception, very few experimental demonstrations impressed me more than the collection of video clips featuring roborat as the leading protagonist. One of these videos depicted roborat climbing a rubber mesh, peppered with holes the size of roborat's paws, that was installed almost perpendicular to the ground. Another showed roborat successfully navigating the entire sequence of demanding obstacles that formed the open-field testing track in San Antonio, Texas, that DARPA used at the time to evaluate the limits of the most advanced autonomous robots being built.

At this point, you may be wondering what could be so sensational about a particular robot performing these tricks. After all, Mitsuo Kawato and Gordon Cheng's robots could sing, dance, and play Ping-Pong, couldn't they? While that was true, those robots had been *programmed* to carry out specific tasks; none of them could sing new songs or shuffle a samba or defeat a Ping-Pong gold medalist without the aid of a roboticist's diligent code. And that certainly held true for finishing the full extent of the DARPA track's obstacles. An autonomous robot might get stuck in the sand trap, ambushed on the rubble pile, or stymied by the very steep and equally slippery ramp.

More than anything, though, roborat's performance irked a number of rival roboticists for the simple fact that it was not a robot at all. It was just a rat, not of the Philadelphian sewer variety, but a proud member of the Long-Evans strain, raised, groomed, and slightly enhanced in John Chapin's lab at the medical school of the State University of New York, Brooklyn. When Chapin entered the DARPA robot track carrying an animal cage, a few people chuckled in disbelief. More of them chuckled when from that cage he carefully pulled out, not a gizmo, but a black-hooded rodent. After spending a few minutes gently patting the rat's back in some kind of Zenlike maneuver, he carefully positioned the animal at the track's starting line.

When Chapin stepped away from his pupil, everyone who was paying attention could now see that his hands were tightly grasping a commonplace laptop. Suddenly, all chuckling and mocking ceased. Chapin's experiment, it should be said, went much further than anything his illustrious Spanish predecessor had accomplished. For several months, he had been testing an elaborate new electrical brain stimulation paradigm. Rather than merely attempting to block or induce sporadic body movements by stimulating a particular brain region, Chapin planned to use electrical pulses to instruct his rat on how to steer through a complex maze. To deliver this kind of instruction, he chronically implanted a single stimulating electrode in a cortical region he knew very well: the whisker representation of the primary somatosensory cortex. Since these stimulating electrodes would be used to inform the rat which direction it should turn its body, Chapin implanted one electrode in the right S1 cortex and one electrode in the left. Different from most previous studies that utilized electrical stimulation of S1, Chapin also implanted a set of stimulating wires in the medial forebrain bundle (MFB), which he knew would produce a very intense pleasant sensation when it was stimulated. After the rat recovered from the implantation procedure, Chapin rigged the animal with a "backpack" that could receive commands from a straightforward radio transmitter and relay a tiny electrical pulse to any one of the rat's implants.

The secret, though, was in how Chapin had defined the temporal sequence in which these pulses were delivered, either to each cortical hemisphere or to the MFB. Basically, he had discovered that his rats could learn to associate a pulse delivered to the right S1 cortex with an instruction to turn right, whereas a pulse delivered to the left S1 cortex signaled an instruction to turn left. Such learning was possible, and rather quick, because every time the rat followed Chapin's instruction correctly, it received a single electrical pulse to its MFB. Using this strategy, he managed to train his roborats to navigate any maze they faced—becoming the first hybrid living thing to smash the DARPA testing track's record.

With his usual modesty and candor, even after such a spectacular performance, John Chapin recognized pretty well the limits of his otherwise

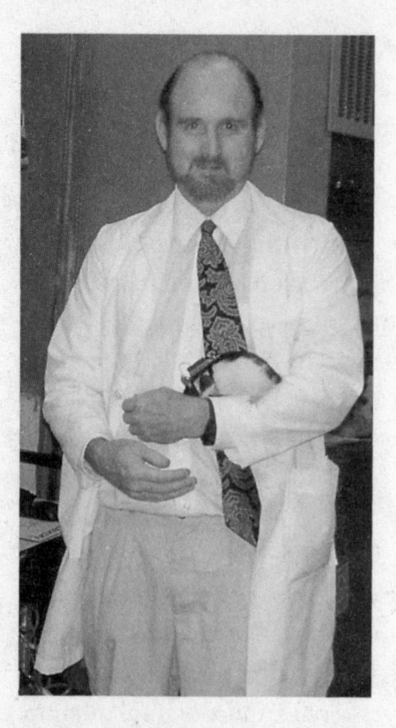

FIGURE 10.2 John Chapin and roborat. In the left panel, Dr. John K. Chapin and his creation. Below, roborat walks over a metal mesh. *(Courtesy of Dr. John Chapin.)*

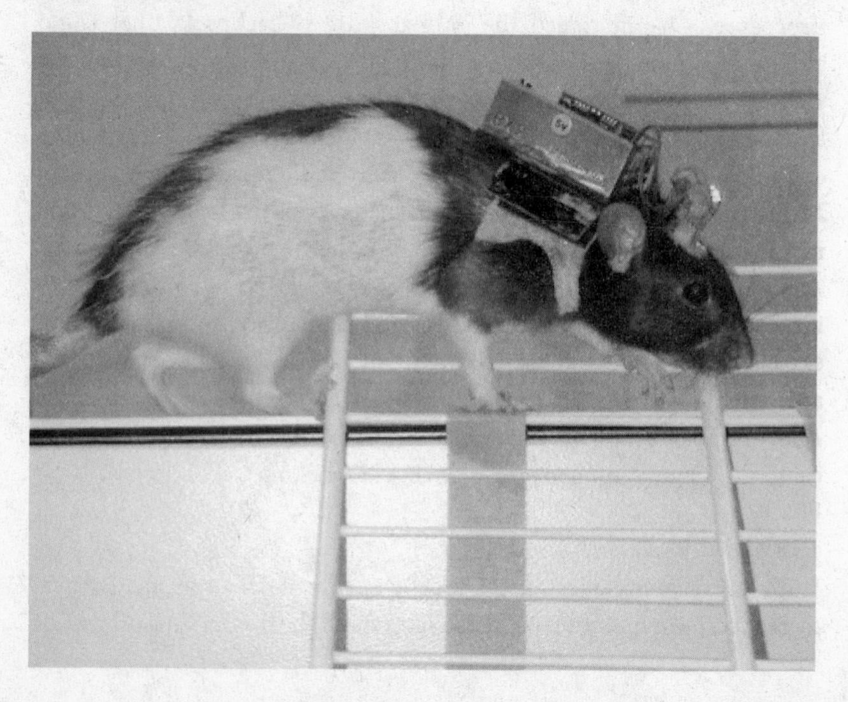

unique strategy to communicate with his subject's brain and reward it for a job well done. Thus, despite later introducing an instruction for forward movement, among other innovations, he was fully aware that adapting his approach in an upper- or lower-limb BMI would require many more commands, and far more nuanced patterns of electrical stimulation, to be delivered to the somatosensory cortex. Regardless, the incredible adventures of roborat helped solidify my hunch that electrical brain stimulation might be used in a next generation of BMIs. It was time to dislodge these conjectures from the drawing board and try them out in a series of experiments.

Before reaching for the most fanciful of these ideas, I decided to attempt something relatively easy: could a monkey learn to interpret a binary message delivered by direct cortical electrical stimulation and use the information acquired to solve a behavioral task? This question grew out of a project that had been recently concluded by Aaron Sandler, a graduate student in my lab, who had demonstrated that owl monkeys could utilize tactile stimuli, delivered to the skin of either their left or right forearm, to identify which of two boxes, placed in front of them, they needed to search for food once an opaque Plexiglas door, which blocked their access, was raised. The two monkeys he trained had no trouble in associating a stimulus on the right arm with finding a food treat in the right box and a stimulus on the left with a treat in the left one. Moreover, after learning this first rule, the monkeys could easily handle a reversal, with a left-arm stimulus meaning the food was on the right and a right-arm stimulus meaning it was on the left.

Having this useful information available, as well as detailed behavioral data that Sandler had carefully gathered during his food-retrieval experiments, Nathan Fitzsimmons started to explore ways of delivering an electrical stimulus to the S1 cortex of the same monkeys (see Fig. 10.3). Sandler had implanted multiple microwire arrays in several cortical areas of his two owl monkeys. This had allowed both Sandler and Fitzsimmons to obtain chronic simultaneous recordings of the electrical activity of close to one hundred cortical neurons for six years in a row. For his thesis project, Fitzsimmons selected a few microwires in the S1 cortex of each animal to deliver electrical cues to the monkeys in a new version of the box-identification task. Moreover, he decided to try out a

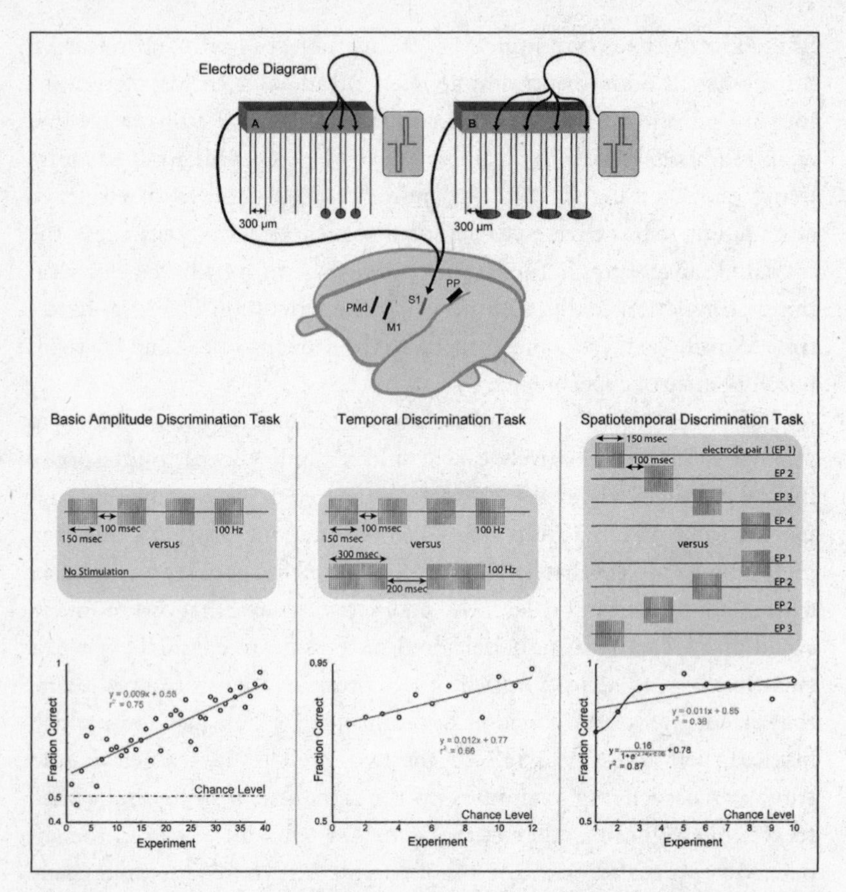

FIGURE 10.3 Talking back to the brain. Top shelf illustrates the experimental paradigm utilized to deliver "electrical messages" to the brain of a monkey. Chronically implanted microelectrode arrays are employed to deliver spatiotemporal electrical patterns that represent different messages. Middle shelf illustrates the different types of patterns utilized to deliver messages to the primate brain: basic amplitude discrimination, temporal discrimination, or spatiotemporal discrimination. Lower shelf illustrates learning curves for each of the three methods to deliver messages. *(Adapted from N. Fitzsimmons, W. Drake, T. Hanson, M. Lebedev, and M.A.L. Nicolelis, "Primate Reaching Cued by Multichannel Spatiotemporal Cortical Microstimulation."* Journal of Neuroscience *27 [2007]: 5593–602.)*

larger variety of "encoding schemes" for delivering the spatial cues directly to the monkey's cortex. His first scheme followed a simple rule: if a high-frequency electrical microstimulation was delivered to the right S1 in the period before the Plexiglas door was raised, the food pellet would be found in the right box, and if no electrical stimulation occurred during that period, the food would be delivered in the left box.

Although the same two monkeys had been extensively trained in the tactile-stimulation version of the task, it took both animals about forty days to learn this new rule to retrieve their food. That was odd. No previous study with cortical electrical microstimulation had ever reported such a long learning period for mastering such basic electrical cues. Still, once the monkeys picked up the association, they performed the task as well as they had while receiving tactile stimulation of their forearm skin. Eagerly, we moved to the next step, which consisted of reversing the rule the animals had just mastered. Now, the monkeys took much less time to learn the cues—about fifteen days. It was time to switch coding schemes again. For this round, Fitzsimmons selected two distinct temporal patterns of electrical microstimulation to transmit the two messages: when the monkeys received an electrical stimulus formed by 150-millisecond-long pulses separated by 100-millisecond pauses, their goodies could be collected in the right box; when they received 300-millisecond-long pulses separated by 200-millisecond pauses, the food was in the left one. With the exception of these small differences in frequency, in this *discrimination contingency* all other features of the stimulus—including the total charge, duration, and cortical location—were identical. Though this was the first time the two monkeys had to discriminate between two slightly different stimuli to select which box they should attend to in order to collect a food pellet, they mastered the new rule in about a week.

Having gone this far, we decided to push the limits of instructing primate behavior with cortical electrical microstimulation. Instead of continuing to employ a single microwire to deliver the electrical pulses, Fitzsimmons enlisted four pairs of neighboring microelectrodes. Then, instead of using distinct frequencies to communicate distinct cues, he chose two traveling waves of electricity, moving in opposite directions across the microwires, to notify the monkeys of where the food pellet had been placed. We suspected our two monkeys would have a much harder time in discriminating between the two patterns in this *spatiotemporal contingency.*

Our concerns proved to be totally unnecessary. In a matter of three to four days, both owl monkeys figured out the subtle distinction between the two spatiotemporal cues and could find their coveted food pellets with the same level of assurance they had attained when facing the three previous coding schemes. They were actually getting faster at learning the cues, as well. It seemed that once the monkeys got the gist of the message we were trying to convey to them—that is, the location of the pellet—they continuously improved their capability of applying this general rule to any new contingency we threw their way.

By identifying a clever way to block the electrical noise created by his stimulator, Fitzsimmons also managed to obtain simultaneous neuronal recordings from the S1 and M1 cortices of his two monkeys around the time the monkeys were receiving the electrical cue revealing which of the two boxes contained the food pellet. When he fed these highly valuable neuronal signals through the same multilinear regression algorithms we regularly used to drive our BMIs, he noticed something interesting. Looking at the linearly combined neuronal activity produced by S1 neurons during the period between the end of the electrical stimulation and the beginning of the animal's movement toward the box, he could predict, solely based on this electrophysiological data, which stimulus had been delivered to the monkey somatosensory cortex just a couple of hundred milliseconds earlier. Encouraged, he fed the activity produced by all recorded M1 neurons during the same trial period into the linear model. This allowed him to predict, with equal precision, to which box the monkey would move its hand, before any sign of movement could be measured in the animal's muscles. As the monkeys mastered each of the task contingencies, Fitzsimmons's predictions also improved. And that was how he was able to document, in great neurophysiological detail, the temporal sequence through which these primate brains decoded the messages embedded in an exogenous electrical stimulus and then seamlessly transformed that privileged information into a decisive act of voluntary will.

Our attempt to establish a direct dialogue with a primate brain had found "ready to listen" neuronal ears. Without a doubt, this was an auspicious start.

By the time Fitzsimmons's results with cortical microstimulation were published, Joey O'Doherty had devised a new series of experiments with rhesus monkeys that utilized cortical electrical stimulation with an upper-limb BMI. O'Doherty and I realized these monkeys could become the first subjects to test a radical new paradigm, baptized in honor of José Delgado as a brain-machine-brain interface. This BMBI would allow the subjects, our monkeys, to interact with a particular artificial device through a closed-control loop that precluded any interference from the body whatsoever—neither the efferent (motor control) nor the afferent (sensory instruction or feedback) components of the interface would rely on any cellular tissue other than the small sample of cortex that was sending and receiving information. The monkeys would receive instructions or sensory feedback directly to their brains, with no need to involve the biological sensors or peripheral neural pathways that primates normally rely upon to gather information from their moving body.

Certainly this was a steep ladder to climb, particularly considering that the first version of such a BMBI afforded a rather impoverished makeshift channel of communication as a substitute for the body's vast sensory apparatus. As in Fitzsimmons's experiments, replacing the senses would require chronically implanted microwires to deliver simple patterns of electrical stimulation, to either the primary somatosensory or the posterior parietal cortex of the monkeys. Our goal was to test whether these monkeys could learn to maximize this single "artificial sensory channel" to decode instructions using their brain activity alone. Later, we would employ the same strategy to give animals feedback about the movements of the machine they were controlling.

To make things a bit less challenging, our first trials were limited to a binary movement direction instruction, which a monkey had to use to determine whether to move a computer cursor to a target on the left or the right of a screen placed in front of the animal. During the early phases of training, the monkey indicated the direction it extracted from the pattern of microstimulation by manipulating a joystick that controlled the cursor's movement. When the animal became highly proficient in executing this maneuver, we gradually removed the joystick and shifted control to the BMI component of our apparatus, so that the animal controlled the cursor's movement with brain activity alone. Typically, the monkey started each attempt by centering the cursor on a starting point, projected in the middle of the screen. Once the cursor was centered, two visually identical circular targets appeared on the monitor, each an equal distance from the cursor's starting point. Simultaneously, an electrical stimulus, which we had coded to represent either the left or right target, was delivered directly to the monkey's S1 cortex. The monkey then had to interpret the microstimulation and generate the pattern of motor brain activity that moved the cursor to the appropriate target (and delivered its truly juicy reward).

Like Fitzsimmons and Sandler before him, O'Doherty launched his experiments by measuring how long the same monkey took to solve the same task when an instruction was delivered to the skin on the animal's arm rather than through cortical microstimulation. These control experiments allowed us to compare the skin and the brain as conduits for receiving the binary instructions needed by the monkey to solve the task. Furthermore, O'Doherty had set up the experiment so that one of the monkeys received electrical messages directly into the S1 cortex, while the other received them into the PP cortex, just a few millimeters behind the S1. This meant we could also compare which cortical region was a better target for guiding the animal's decision through cortical microstimulation.

Similar to what we had experienced during our experiments with owl monkeys, it initially took a few weeks of training for the monkeys to learn the task when the instructions were delivered tactilely. After this start-up period, a sharp divide emerged between their performances based on whether the monkey was slated for S1-directed messages or PP-directed ones. The monkey receiving microstimulation to the S1 cortex soon reached the same level of proficiency in selecting between the two targets as it had achieved with a tactile stimulus. Things were murkier with the other monkey. Although this animal also learned to decode the information delivered by tactile stimulation of the skin, it did not get a handle on how to interpret the electrical messages delivered to the PP cortex. Though we cannot rule out the possibility that either different types of electrical signals or longer training periods are necessary for monkeys to learn how to handle electrical instructions delivered to the posterior parietal region, O'Doherty's experiment opened several scenarios for applications in future BMBIs.

Our team at Duke now turned to the possibilities for establishing a direct communication channel to deliver novel messages to a living brain. Long before my phone conversation with Luis Baccalá in 2005, I had recognized that these new technologies promised to take us far beyond a strict dialogue with the nervous system of our laboratory animals—indeed, they could reshape the brain's own point of view—based on our accumulated body of research. This concept, which I have introduced throughout this book, is what I call the *relativistic brain hypothesis.*

THE RELATIVISTIC BRAIN HYPOTHESIS

When faced with new ways to obtain information about the statistics of the surrounding world, a subject's brain will readily assimilate those statistics, as well as the sensors or tools utilized to gather them. As a result, the brain will generate a new model of the world, a new simulation of the subject's body, and a new set of boundaries or constraints that define the individual's perception of reality and sense of self. This new brain model will then continue to be tested and reshaped throughout the subject's life. Since the total amount of energy the brain consumes and the maximal velocity of neuronal firing are both fixed, it appears that neuronal space and time would have to be relativized according to these constraints.

To test this hypothesis, we have been designing two experimental paradigms that have never before been tried in brain research. The first of these experiments, still in its early stages, entails creating, on a small scale, a

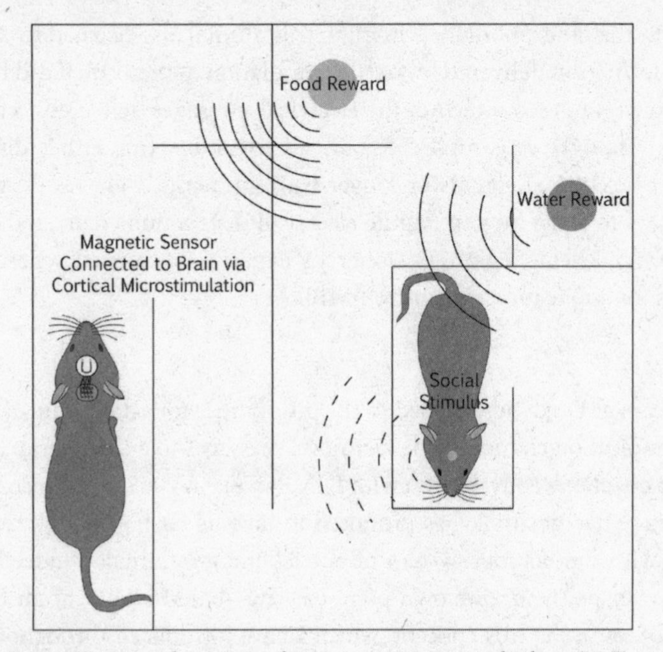

FIGURE 10.4 Drawing depicting a future experiment in which an R6-T rat will be implanted with a magnetic field sensor that delivers electrical microstimulation to the animal's primary somatosensory cortex proportional to magnetic fields of different magnitudes. Each of these different magnetic fields identifies different objects, things like food, water, and the location of a toy rat. *(Illustrated by Dr. Nathan Fitzsimmons, Duke University.)*

"new world"—a very unusual, far from natural environment where newly immersed subjects, in this case adult rats, had never been, let alone lived, in their whole existence. We have chosen to establish a "magnetic" world, a place where all relevant environmental features—things like the world's overall spatial boundaries; the location of food, water, and noxious stimulus sources; the places for nesting and socializing with other animals; and the site where predators (and fear) reside—are indicated by different magnetic field sources (see Fig. 10.4). To investigate whether subjects can learn to navigate in such an environment, a small magnetic sensor will be implanted on the frontal bone of the rat's skull. When the "augmented" rat visits the new world, the implanted sensor will signal the nearby presence of a particular magnetic source by triggering the delivery of a unique spatiotemporal electrical stimulation to the animal's S1 cortex via chronically implanted microwires.

As the rat explores the magnetic world, it will receive two sets of rewards every time it correctly identifies a spot containing a positive stimulus: the "natural" reward (food, water, or social interaction) and an extra incentive (an electrical pulse delivered to the MFB, the brain structure utilized by John Chapin to train his roborats). Conversely, each time a rat mistakenly enters the "predator's compartment" or another negative stimulus, a loud alarm will reprimand the careless wanderer. To ensure that, from moment to moment, the rat will only navigate the environment using the unique magnetic signature of each spot, the locations of the hedonic and unpleasant areas will be rotated frequently. Careful control experiments will also be performed to rule out the possibility that rats detect other sensory cues to the location of rewarding places.

We have named our subjects the "sixth-sense magnetic rats," or R6-T, the T representing the tesla, the unit employed by the International System of Units to designate the strength of a magnetic field. As we build this magnetic world and prepare to introduce the first R6-T to its new home, there are many interesting questions that come to mind. Will R6-Ts be able to learn to interpret the magnetic messages delivered to their S1 cortex? Will the animals learn to live in this totally unnatural world and rely solely on their newly acquired magnetic sense to find food and water and avoid predators and other awful encounters, while guiding themselves to their housing quarters or the company of other rats? And if R6-Ts can learn all these things, will a full representation of the magnetic world emerge in their S1?

My prediction is that the R6-Ts will eventually be able to learn some, if not all, of the crucial parameters of this brave new magnetic world, and that clear magnetic receptive fields will emerge to supplement the rats' typical tactile responses. I am also confident that these magnetic RFs will not impair the ability of R6-Ts to use their whiskers to discriminate the normal range of tactile stimuli that all regular rats normally perceive. In the context of the relativistic brain, this will be possible because the animals will incorporate the statistics of the magnetic world on top of those that describe their natural environment, the one that has been assimilated in their brains since early postnatal life.

It is important to emphasize that there is nothing special about selecting different magnetic sources for building the new-world environment of these experiments. As a matter of fact, if my theory is correct, one would obtain the same results if the experiments were carried out in an "infrared" or in a "ultrasound" world. Recent research further suggests that electrical stimulation could also be replaced as the method of choice to deliver environmental information to our rats' brains. A very good candidate for such a job is optogenetics, a revolutionary new method introduced in 2006 by Karl Deisseroth, an associate professor of bioengineering and psychiatry at Stanford University. In optogenetics, a light stimulus is used to modulate the electrical activity of populations of cortical neurons. This isn't as simple as flashing strobe lights, however. First, the cortical neurons need to be infected with a virus that carries the genetic information needed for the synthesis of particular proteins that form unique ion channels that respond to certain wavelengths of light. For instance, if a cortical neuron has been genetically engineered to express Channelrhodopsin-2—a protein that is known to direct movement in response to light in green algae—a light stimulus of a distinct wavelength will open ChR-2's sodium channel, triggering a massive influx of sodium ions into the neuron and making it fire an action potential. Conversely, if the cortical neuron has been instructed to produce a light-gated chlorine channel, using a different light-sensitive protein, another stimulus of a different wavelength will inhibit this cell's firing. By mixing up these two types of light-gated ion channels in the S1 cortex of our rats, we could map each of the magnetic signatures found in the magnetic world with a pattern of light stimuli that generates a particular pattern of cortical electrical activity. The great advantage of using optogenetics in such studies is that it does not produce artificial electrical artifacts and does not cause tissue damage.

Setting these technical considerations aside, I envision that long-term exposure to a magnetic world will not only lead to a seamless incorporation of the magnetic field sensor into the R6-Ts' brains, but also to the emergence of a new sense of worldly and bodily reality in these animals' minds. Admittedly, it would be rather difficult to prove this in an animal that cannot express its views, let alone its thoughts,

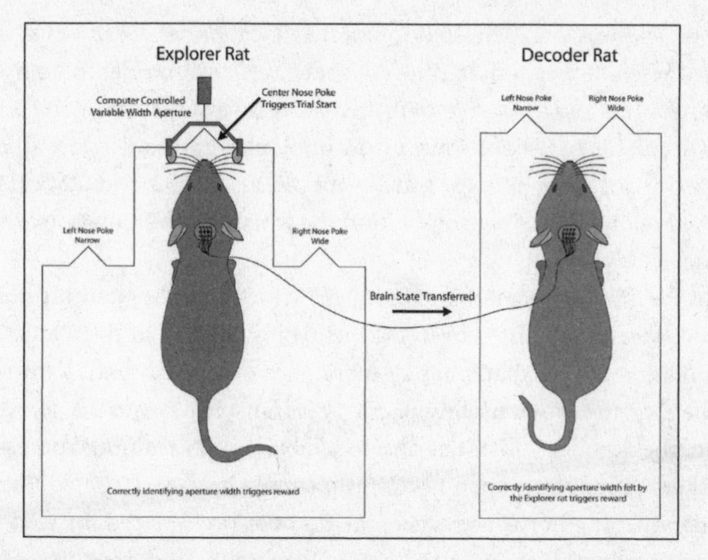

FIGURE 10.5 The explorer rat and the decoder rat. Illustration of a true brain-to-brain interface linking an explorer rat, which uses its facial whiskers to discriminate the diameter of a variable aperture, with a decoder rat whose main function is to indicate the diameter of the aperture based on the pattern of brain activity transmitted by the explorer rat, without ever touching the aperture with its whiskers. The brain-to-brain interface connects the brains of the two rats. *(Illustrated by Dr. Nathan Fitzsimmons, Duke University.)*

with rhetorical vehemence. Yet, these predictions follow logically from the experiments we have conducted thus far.

The second of our experiments is focused on testing the first brain-to-brain interface (BTBI). In our preliminary design, a rat previously implanted with arrays of microwires in the whisker representation of the S1 cortex will be trained to use its long facial vibrissae to perform the tactile discrimination task that Eshe faced in chapter 5. Once the rat has learned how to discriminate the relative width of two openings, judging one as "narrow" and the other as "wide," it will be promoted to "explorer rat" status (see Fig. 10.5). At this point, the explorer rat's patterns of neuronal electrical activity, recorded from the whisker representation of S1, will be wirelessly transmitted to another location, a closed behavioral box, where another animal, the "decoder rat," will reside, in complete darkness, just waiting. We will then activate a multichannel electrical stimulator or a light source grid implanted in the decoder's head, triggering a spatiotemporal wave of electrical or light stimulation targeting populations of neurons within the whisker representation area of the decoder's S1.

In the first version of this experiment, which is in the planning phase, the decoder rat will have been independently trained in the tactile discrimination task so that it has a general idea of the task goals. However, by the time the explorer's brain activity is transported into the decoder's brain, the decoder will not be able to fall back on its training and use its whiskers to judge an opening's width, simply because there will be no similar opening in the box where the decoder rat is placed. Instead, the lonely decoder will have to indicate behaviorally, by poking its nose at one of two spots in the box's walls, whether the opening sensed by the explorer rat's whiskers was a wide or a narrow one. The decoder rat will have to rely solely on its own brain's transduction of a fragment of the electrophysiological correlates of the explorer rat's tactile experience to make a judgment about the diameter of an opening that its own whiskers have never brushed. To make things more interesting, every time the decoder reports the correct diameter of the opening and gets rewarded for a job well done, the explorer rat will receive an extra amount of reward for successfully transmitting its perceptual experience to its partner.

Clearly, there are many potential problems that might derail the execution of this complex experiment. But assuming we can settle all the technical details, and both animals are capable of learning how to interact virtually with each other, there will be many implications for testing and developing more sophisticated BTBIs. My prediction is that individual S1 neurons in the decoder's brain will become responsive to any mechanical stimulation of the explorer's whisker, even when this pair of rats is not engaged in a proper tactile discrimination task. For instance, I expect that the decoder rat's tactile receptive fields will expand to include not only its own facial whiskers, but also the whiskers belonging to the explorer rat. If this happens, it will indicate that even a

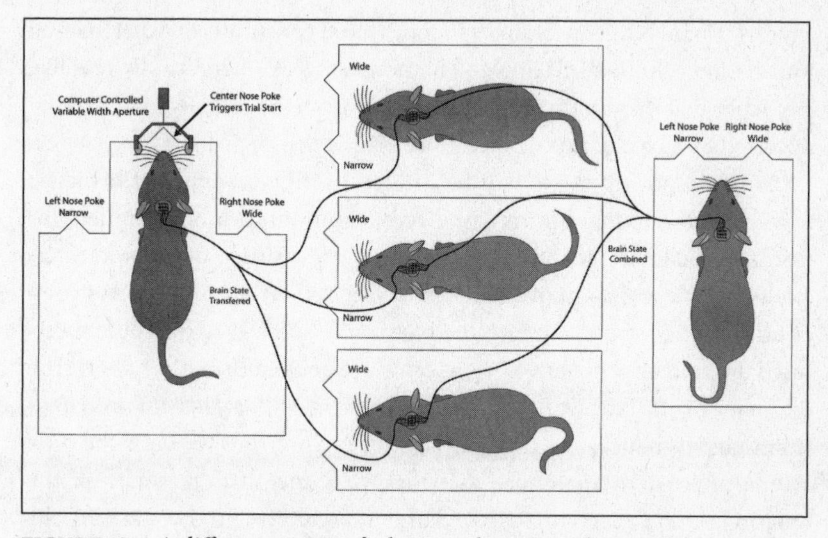

FIGURE 10.6 A different version of a brain-to-brain interface involving an intermediary layer of rats between the explorer and decoder rats. *(Illustrated by Dr. Nathan Fitzsimmons, Duke University.)*

rudimentary BTBI will extend the brain's internal representation of the body to include the body of the other brain connected to it.

Such a finding would be nothing short of astounding. It would be the first successful demonstration that two living brains can be functionally linked and, as a result of this communion, collaborate harmoniously toward the completion of a mutually beneficial goal.

So far, however, I have only considered the simplest of the imaginable BTBIs, one in which the brain of one subject, the explorer, communicates with the brain of another subject, the decoder, in a unidirectional manner. Yet, the drawings I sent to Luis Baccalá in 2005 contained a number of variations on this original scheme, one of which is illustrated in Figure 10.6. What if the decoder rat is granted the right to send its own brain activity back to the explorer rat? In this arrangement, could the two brains eventually reach a consensus, let's say, about the identity of a complex object explored only partially by each rat? Would the rats literally share their minds to build vicarious sensations, through a sort of touchless, Vulcan mind-melding ritual, in order to overcome the limits of their individual brains? Considering current state-of-the-art BMI technology, such a bidirectional brain-to-brain exchange would constitute an experimental tour de force. Yet its full implementation is at least theoretically feasible, particularly if our unidirectional BTBI proves to be successful and easily operated by a pair of rats or, down the road, a pair of primates.

But I do not think we will be limited to the possibility of bringing together two living brains in direct communication. Imagine, for instance, that instead of being broadcast to a single decoder rat, the explorer rat's brain activity finds its way to each of the brains of a group of animals, identified as "intermediary rats" (see Fig. 10.6). These intermediary rats also do not have access to direct information about the diameter of the real-world opening. They can only gather the information that is broadcast by the brain of the explorer rat, which is the only animal that can actually use its whiskers to measure the width of the opening. Upon receiving the explorer rat's brain activity, each of the intermediary rats must decide to take a particular course of action: poking its nose to either the left or the right to indicate whether it believes the pattern of electrical or light messages indicates the detection of a narrow or a wide opening. As the intermediary rats make their individual choices, neuronal electrical activity from each of their brains is then broadcast to a decoder rat, whose new job is to evaluate the collected "brain opinions" of the intermediary animals and determine if the opening is narrow or wide. As fair payment for an extraordinary job collectively achieved, the pack of intermediary rats and the curious explorer would each receive a sweet reward, delivered promptly every time the decoder identifies the correct diameter.

That would surely be a stunning adaptation, a true consensus formed out of the virtually experienced, yet preciously unique, fleeting traces of a communal electrical brainstorm.

11

THE MONSTER HIDDEN IN THE BRAIN

In different periods of my life, I have looked at the brain with different eyes. For instance, back in middle school, I thought the brain resembled a highly elaborate supercomputer, one so complicated and mysterious that not even the great Mr. Spock, of the *Starship Enterprise*, could hope to grasp it. I reasoned that if the Vulcan scientist who once intoned, "I am endeavoring, ma'am, to construct a mnemonic circuit using stone knives and bearskins," was often mystified by the logic employed by the human brain, the whole endeavor of understanding how that brain actually worked was pretty much fruitless. Intimidated, I dropped the subject and focused instead on sharpening my defensive midfield soccer skills, dreaming that one day I could play for Palmeiras. Unfortunately, that first career option did not pay off as intended.

By high school, a casual and life-changing encounter with Isaac Asimov's 1964 book *The Human Brain* rekindled my interest. On each page, as Asimov related the macroscopic and microscopic structures that formed the brain, my astonishment grew. But I could not believe, when I reached the end, that there was no chapter that actually clarified how these structures, with their quaint Latin and Greek names, toiled together.

Just a few years later, in medical school, I literally dissected the brain, to which Asimov had introduced me, along many competing layers. Depending on the class you attended, the brain was undressed according to an anatomical, histological, physiological, pharmacological, neurological, or psychiatric point of view. Once again, no one came forward, not even my scientific hero, Dr. César Timo-Iaria, to tell us how all these levels of organization assembled themselves into thinking. At the time, to be a legitimate neuroscientist, you had to become an expert in one—and only one—of these disciplines, and dedicate your life to working within the boundaries of the accepted canon of your chosen field.

After graduating from medical school, I decided to pursue a PhD in physiology rather than apply for medical residency. With the total support of Timo-Iaria, however, I embarked on the risky research project that I described in chapter 1: using graph theory and computer programs to analyze a brain circuit. Halfway through my thesis project, the American cognitive scientist Marvin Minsky published his famous treatise *The Society of Mind*. Eagerly anticipating that the leading figure in the field of artificial intelligence had found the ultimate answer, I dove into the book, only to discover that what I sought could not be found there either. Minsky did not seem to care much about real brains, just about the higher-order operational processes that may take place inside them.

Close to the time of my thesis defense, I attended a lecture by the American physicist John Hopfield. In the mid-1980s, Hopfield had invented associative artificial neural networks, and I remember being fascinated as we talked for almost an hour about his opinion of what the future held for neuroscience. That was when I became interested in the notion that I could only vaguely define then as "neural dynamics." Later, the conceptualization and implementation of brain-machine interfaces emerged as my attempt to search for the physiological principles that govern the dynamic operation of vast ensembles of neurons in freely behaving animals.

For the past few years, deeply influenced by the findings of two decades of research effort, I have found myself looking at the brain in a very different way. Somewhat surprisingly, even for me, today I like to compare the brain to a special type of ocean, a never-idle sea of electricity, held together by multiple synchronous waves of neuronal time and capable of remembering everything that sails through its mysterious gray waters. Using this liquid metaphor, I mentally visualize how the true biological substrate of a relativistic brain actually works. Just as someone who is interested in understanding the behavior of ocean currents, whirlpools, maelstroms, and tsunamis would be unable to explain these macroscopic phenomena by analyzing the behavior of water molecules or atoms, focusing exclusively on the properties of single neurons becomes a major distraction if one wants to understand the behavior of the whole neuronal ocean. Given this new metaphor for the brain, it is only appropriate that some of my recent findings in understanding thinking began with the unforgettable moment my students and I spotted a mythological monster deeply buried in the meandering patterns of mammalian brain dynamics.

"It looks like the Loch Ness monster!" Improbable as it sounded, that was exactly the image that came to my mind when I saw the strange, two-dimensional distribution plot of electrical activity in the brains of rats that had just been handed to me.

"How in heaven can you see a fake Scottish dragon in this data?" A few inevitable giggles surfaced from the lab's benches.

"Look at the plot. On the top right corner you can almost see the monster's elongated, triangular head. The head is connected to a very long neck, like a plesiosaur's, that runs diagonally. Then the neck coalesces into a massive elliptical body, right at the plot's center. And from the bottom of the body you can clearly see what looks like hind paws—adapted for swimming!" Even I was amazed by how readily my brain assigned all these anatomical parts to a picture that, in truth, contained a bunch of little black dots.

"I was thinking about calling it the Great Brain Attractor." Shih-Chieh "CJ" Lin, a Taiwanese graduate student in the lab now a researcher at NIH, had clearly thought long and hard about what to name the main product of his months of arduous mathematical toil alongside Damien Gervasoni, a French postdoctoral neuroscientist from the city of Lyon, and Sidarta Ribeiro, who, in his doctoral thesis at Rockefeller University, had categorically implicated sleep as a key contributor to memory consolidation.

Upon joining my laboratory, Gervasoni and Ribeiro had decided to employ our approach for chronic, multisite, multi-electrode recordings to investigate the normal wake-sleep cycle of freely behaving rats. First, they implanted microwire arrays in four distinct brain structures: the somatosensory cortex, the thalamus, the hippocampus, and the caudate-putamen complex, an important component of the basal ganglia. After the animals recovered from that procedure, they habituated each rat to live in a comfortable experimental chamber where it could have access to food and water with liberty. Each rat spent up to five days in this cozy spot, on a twelve-hour day/night cycle, doing everything regular laboratory rats do as they go about their lives. Meanwhile, Gervasoni and Ribeiro aimed two high-resolution video cameras on the rat's chamber, continuously monitoring, *Big Brother*–style, every detail of behavior. In addition, they recorded the rat's neural activity, in the form of multiple local field potentials (LFPs)—the sum of electrical signals from a chunk of tissue—from each of the four implanted brain areas. The LFPs and video recordings were synchronized to within milliseconds. Since LFP activity tends to encompass a very large pool of neurons in any given brain location, Gervasoni and Ribeiro took very extensive recordings, up to ninety-six hours, of large-scale cortical and subcortical activity, which they could then compare to the variety of behaviors produced by the rats as they shuffled through hundreds of wake-sleep cycles. They also recorded the electrical activity of tens of single neurons widely spread around each of the four structures from which they sampled LFPs, as a check.

Interestingly enough, although a vast number of neurophysiological correlates of sleep had been identified by the early 2000s, there was no approach in place for predicting an animal's overt behavior, at each moment of the wake-sleep cycle, based solely on the electrical activity of its brain. For instance, neurophysiologists knew that, during different segments of the wake-sleep cycle, particular types of neural oscillations emerged (see Fig. 11.1). These oscillations are so ubiquitous that neuroscientists have often tried to implicate them in a variety of functions, including a definition of "consciousness." We were a bit more humble in setting up our experiment.

During the wake-sleep cycle, each of the neural oscillations is typically characterized by a distinct amplitude and dominant frequency range. For example, when rats are in a "quiet awake" state—meaning that they are standing still on their four paws, but not producing rhythmic whisker movements—cortical recordings log a low-amplitude, high-frequency oscillation primarily distributed in the beta (ten-to-thirty-hertz) and gamma (thirty-to-eighty-hertz) ranges. These low-voltage, high-frequency oscillations usually define a state of widespread cortical desynchronization. I compare this to a quiet ocean, with no major waves observed on its surface (see Fig. 11.1).

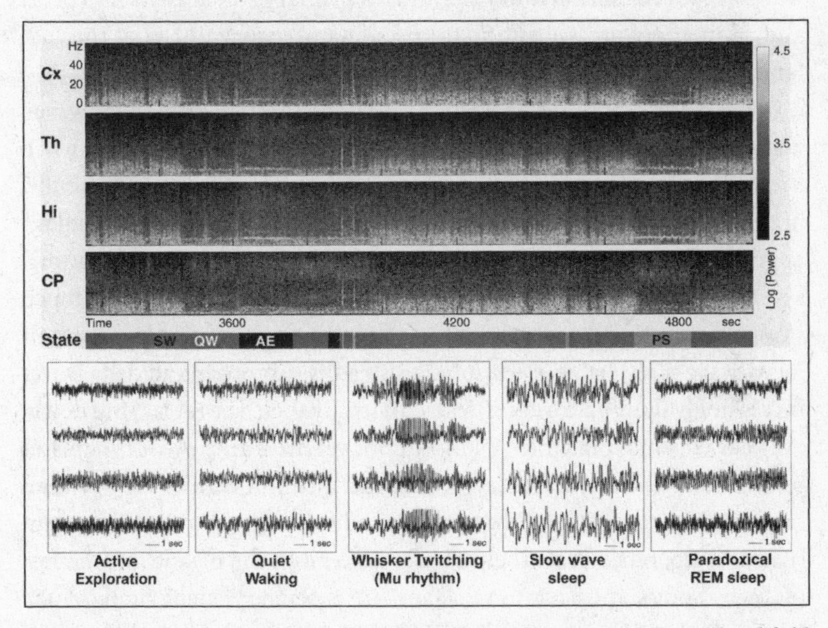

FIGURE 11.1 Spectrograms (top four 3D graphs) and corresponding raw local field potentials (5 graphs in the lower shelf) illustrate the frequency and general time pattern of different types of brain oscillations observed under different types of behaviors: active exploration, quiet waking, whisker twitching, slow wave sleep, and paradoxical REM sleep. In the spectrogram (top shelf) graphs, *X* axis is time (with the period of each behavioral state marked) and the *Y* axis depicts the frequency of oscillations. Gray scale in the *Z* axis depicts the magnitude or power of a particular frequency of oscillation for each state. *(Adapted from D. Gervasoni, Shih-Chieh L., S. Ribeiro, E. S. Soares, J. Pantoja, and M.A.L. Nicolelis, "Global Forebrain Dynamics Predict Rat Behavioral States and Their Transitions."* Journal of Neuroscience *24 [2004]: 11137–47.)*

As rats start moving around to explore their surrounding environment, a state we labeled "active exploration," they sniff around, lick around, and make large-amplitude whisker movements. In their brains, a conspicuous five-to-nine-hertz oscillation rhythm, known as the theta rhythm, arises in the cortical and subcortical structures, in addition to the beta and gamma activity of the quiet-awake state (see Fig. 11.1). As we saw in chapter 5, a distinct pattern of rhythmic cortical firing is evident when awake and immobile rats start producing delicate, low-amplitude "whisker twitching" movements. This whisker-twitching state registers oscillations at seven to twelve hertz, the same frequency as the whisker movements, first in the animal's primary somatosensory cortex and soon afterward spreading to the somatosensory nuclei of the thalamus and other subcortical structures.

This pattern of brain oscillations is suddenly transformed, however, when alert rats descend into the early stages of sleep. At moments when the video cameras capture the animal becoming drowsy, a high-amplitude, highly synchronous neural oscillation, known as "sleep spindles," appears. Classically, sleep spindles are described as waxing and waning, seven-to-twelve-hertz waves of rhythmic activity that ride on top of much slower one-to-four-hertz delta waves. Thus, as a rat plunges deeper into a "slow-wave-sleep" state, sleep spindles steadily disappear and delta waves increasingly dominate the cortical activity. By then, the rat is lying down, eyes closed, enjoying a nice nap—but not yet dreaming. As it progresses into its sleep cycle, the rat starts to dream, albeit in a different way than humans do, as far as we know: rather than displaying rapid-eye-movement (REM) sleep, rats vibrate their facial whiskers! In the meantime, the rest of their body's musculature relaxes, in a serene rodent atony. Such unusual behavioral patterns of REM sleep in rats, which should be more properly designated as "rapid whisker movement" sleep, were originally reported in 1970 by my mentor, Dr. César Timo-Iaria, the first sleep neurophysiologist to suggest that, different from the visually dominated sleep experiences of primates, rats likely delight in rich tactile dreams. That peculiarity notwithstanding, the electrophysiological signature of REM sleep in rats, as in other mammals, is marked by the recurrence of low-amplitude, high-frequency oscillations that are virtually identical to the cortical desynchronization patterns observed during rat wakefulness.

Intriguingly, theta oscillations—the frequency seen in active exploration—appear in the hippocampus during REM sleep.

Though these details of the wake-sleep cycle had been known for some time, no one had devised a way to illustrate this electrophysiological information in a single, two-dimensional graph. Lin succeeded—graphically representing the intrinsic global dynamics of the brain, a central component of the relativistic brain's own point of view. He started by using a well-known algorithm, the fast Fourier transform, to identify the frequency bands in the cortical and subcortical LFP recordings that contributed the most, in terms of power (or amplitude), to the frequency spectrum of the ubiquitous neuronal oscillations that permeate the rat wake-sleep cycle. That analysis yielded four frequency ranges of interest: 0.5 to 20 hertz, 0.5 to 55 hertz, 0.5 to 4.5 hertz, and 0.5 to 9.0 hertz. Next, he tested which potential relationships between these frequency bands could most clearly differentiate between the different brain states exhibited during the wake-sleep cycle. After an exhaustive search, he settled on two spectral amplitude ratios, with one calculating the ratio of 0.5-to-20-hertz oscillations versus 0.5-to-55-hertz and the other the ratio of 0.5-to-4.5-hertz oscillations versus 0.5-to-9-hertz. Since in these ratios the numerator was always part of the frequency range of the denominator, the resulting values conveniently fell between zero and one.

Having identified the appropriate metrics, Lin calculated the two spectral amplitude ratios for every consecutive second of each of the many hours the LFPs were obtained concurrently in the cortex, hippocampus, thalamus, and striatum of the rats. For each animal, four extremely long time series (one for each brain area recorded) were obtained for each ratio, with one value, between zero and one, calculated for each second in time. To generate a two-dimensional plot depicting all this information, Lin took advantage of a multivariate statistics technique to linearly combine the data from the two clusters into a single record. Then it was a simple matter of plotting the values on an *X-Y* "state-space" graph.

The Loch Ness monster I saw during my first encounter with Lin's Global Brain Attractor is shown in Figure 11.2. It is immediately clear that the individual dots are not distributed around the graph in any sort of random pattern. Instead, they define a highly organized image in which a few very dense clusters of dots are separated from one another by much lighter and smeared "belts" of points. When we subsequently analyzed the plot of brain oscillations against the videos, we discovered that each of the dense clusters could be mapped to one of the major behavioral states seen in a normal rat's wake-sleep cycle, and the lighter, smeared "belts" of dots coincided with a rat's transition from one of these states to another. Because each point in the state-space corresponds to one second of brain activity, the denser dot clusters—those that represent the main behavioral states—correspond to where and how the animal's brain spends most of its time. The distribution of time spent in each of the awake-sleep states can be seen in a plot called the hypnomap, shown in Figure 11.3. In contrast, transitions between these states occur much faster, which is why they appear smeared on the graph.

Lin's elegant plotting of state-space opened the way for predicting an animal's general behavior during the wake-sleep cycle from raw electrophysiological records alone. For example, the very dense ellipsoid cluster at the center of the state-space, which resembles the Loch Ness monster's body, depicts the time when the animal is fully awake. If the amplitude of spectrum ratios matches up with a dot that falls into the rightmost two-thirds of the ellipsoid, the rat will very likely be awake, standing on its four paws but not moving at all, not even its whiskers. On the other hand, if the internal brain dynamics produce a dot in the leftmost third of the same ellipsoid, that same rat will very likely now be awake and moving around, exploring its surrounding environment. If the dot appears in what looks like the back of the monster's head, the rat is likely becoming drowsy, ready to begin a nap, as it falls into slow-wave sleep. A more lightly populated "neck" connects the "awake" body to the "sleepy" head since the brain oscillates in this region much less than it does in the other two. That is because the neck represents the dynamic transition between being awake and the early stages of sleep.

Now, if you look closely, you will see a much smaller but still dense cluster of dots next to the monster's body that looks like something of a wing. This smaller ellipsoid is connected to the monster's head by a faint, curvy cloud of dots. It turns out that this small ellipsoid matches

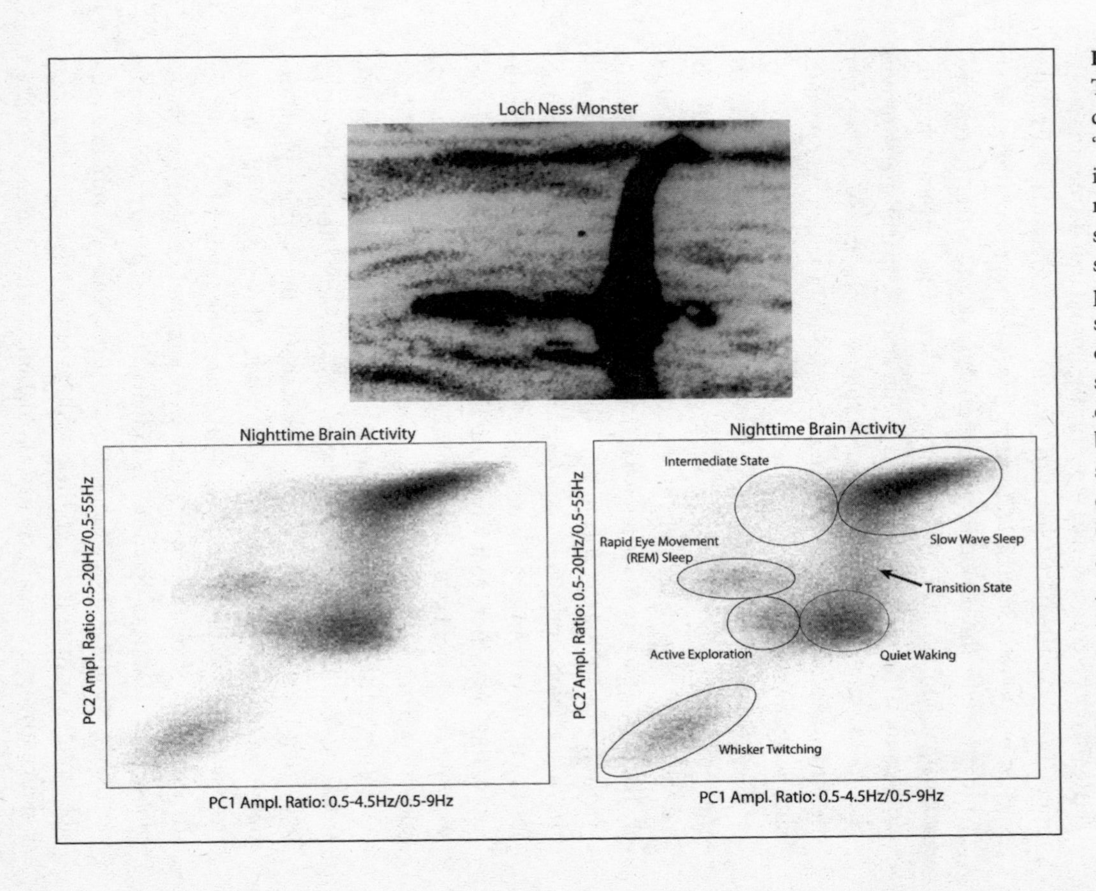

FIGURE 11.2
The Loch Ness Monster, as depicted in the classic "fake" photograph, appears in the top shelf. Its brain relative is depicted in the state-space of the lower left shelf. On the lower right panel, the different brain states that correspond to different locations in the state-space are marked by ellipses. A transition state between quiet awake and slow-wave sleep is also depicted. *(Adapted from D. Gervasoni, Shih-Chieh L., S. Ribeiro, E. S. Soares, J. Pantoja, and M.A.L. Nicolelis, "Global Forebrain Dynamics Predict Rat Behavioral States and Their Transitions."* Journal of Neuroscience *24 [2004]: 11137–47, with permission.)*

up with the time when the rat is experiencing REM sleep, whereas the faint cloud corresponds to the transition between slow-wave sleep and REM sleep, a state that became known as "intermediary sleep." (REM sleep is also connected to the quiet-awake state by a brief and faint transitional state; unfortunately, it is difficult to capture it distinctly in a two-dimensional plot, but it stands out in a three-dimensional rendition.)

Finally, in the lower leftmost quadrant of the state-space, a dense cluster of dots is connected to the central ellipsoid that defines the awake state. These are the dots I liken to the monster's foot, adapted for fast swimming in sweet Scottish lake water. This "foot" represents the whisker-twitching state, the period in which fully awake and attentive rats produce small-amplitude, high-frequency whisker twitches. Meanwhile, the dots connecting the foot to the main ellipsoid—a sparse smear of dots that looks something like a leg—corresponds to the transition between the quiet-awake immobile state and the period in which rats produce such whisker twitches.

Over the past five years, in every rat we have studied, we have mapped the same general Loch Ness–looking wake-sleep state-space. With the exception of the whisker-twitching state, for which the precise position in the state-space varies somewhat across subjects (a problem that is resolved by mapping the dots in three dimensions), the locations of all of the major behavioral states and transitions remain very constant, from rat to rat. Even in mice and monkeys, the general shape is basically the same. Thus, we believe that when data are available for the human brain, we will also see a Loch Ness monster—minus the whisker-twitching state, of course.

To test our findings, we generated other plots by analyzing the data from the four cortical and subcortical regions separately, rather than averaging them together. Even in this scenario, the Scottish predator's body seemed to emerge from the mist of dots. This suggests that no one precise spatial location in the brain—at least in the subdivision called the diencephalon—from which the data were sampled defined the brain's global internal state of dynamics. Ultimately, the wake-sleep cycle can be retrieved from any of the diencephalic structures. And by combining multiple brain structures together, the resolution of the final image increases significantly, appearing almost like a hologram, much as the distinguished

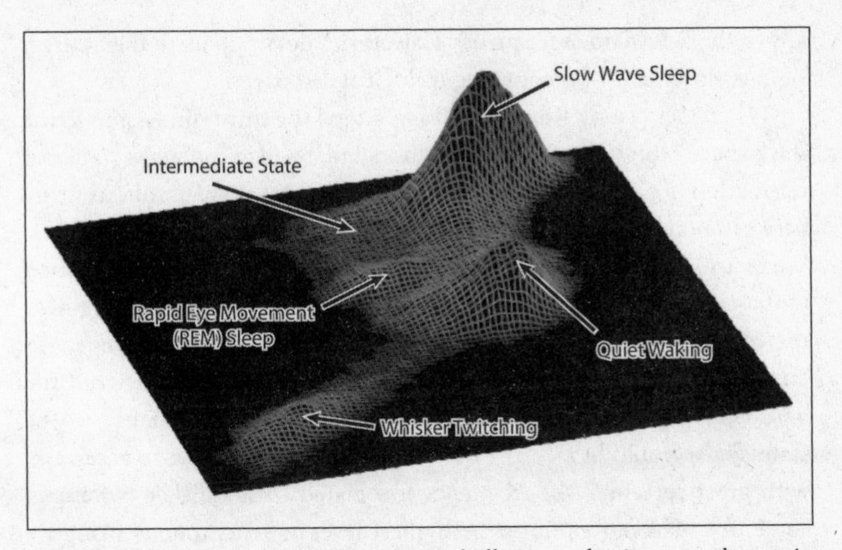

FIGURE 11.3 The hypno-map. This 3-D graph illustrates the time spent by rats in each of their main behavioral states. The *X* and *Y* axes of the graph depict the state-space shown in Figure 11.2, while the gray-scaled *Z* axis represents the time spent by rats in each state. Notice that for most of their lives, our furry friends sleep, without dreaming, in the slow wave sleep state. Quiet awake is only the second most common brain state in the life of rats. *(Courtesy of Dr. Shih-Chieh Lin, National Institute on Aging, NIH; and Dr. Damien Gervasoni, Claude Bernard University, Lyon, France.)*

neurosurgeon and neurophysiologist Karl Pribram, a student of Karl Lashley, imagined it a few decades ago.

Perhaps more stunning, we could create a real-time animation of the Loch Ness state-space, which allowed us to envision how the distribution of oscillation points is "built" in freely behaving animals, mapping the rat's brain electrical activity as the animal shifts, from moment to moment, from one physiological brain state to another. This animated version of the state-space revealed that the rat brain navigates the frequency state-space using highly stereotyped and preferred dynamic trajectories. By the same token, we discovered that there are some trajectories that simply cannot occur. In many ways, these trajectories matched what has long been observed in animal behavior, that animals do not jump, for instance, from whisker twitching or quiet attentiveness into REM sleep, or from active exploration into slow-wave sleep. So, if any of those forbidden frequency trajectories does pop up in the brain, it signals a problem—perhaps a neurological disorder.

When Gervasoni, Ribeiro, and Lin added the third dimension to the state-space plot, it also allowed us to include *pooled coherence*, a global estimation of the second-by-second level of coherent synchronous firing between populations of neurons located across all four brain structures. We focused our estimate on the seven-to-fifty-five-hertz range, and found that as the pooled coherence calculated in a given one-second segment increased, the synchrony in this frequency range became stronger. To our delight, this further enhanced the spatial separation between the dense clusters—those that identified the key behavioral states—in the state-space graph. In fact, the pooled coherence allowed us to ascertain, with great certainty, the frequency associated with whisker twitching, since this behavior exhibits the highest level of synchronous firing we observed across the brain structures we sampled.

This simple modification also revealed that most of the quick transitions between major states were characterized by abrupt, strong changes in firing synchrony. For example, when the rat was shifting from the quiet-awake state to slow-wave sleep, or from slow-wave sleep to REM sleep, the brain structures all appeared to start firing more synchronously at the sleeping-spindle frequency range of seven to twenty hertz. In the case of the transition to REM sleep, this abrupt increase in coherence is defined as the state of intermediary sleep. The animation of the state-space also showed that trajectories between a "departing" state and a "target" state (e.g., from attentive wakefulness to slow-wave sleep) could fail, making the brain activity return to the departing state. We later realized that these failures occurred primarily because the level of coherent synchronous firing did not reach the threshold necessary for going all the way to the target state.

Over the past half century, multiple laboratories have found evidence that the transitions between different states in the wake-sleep cycle are determined by the interplay of a series of modulator neurotransmitters. These chemicals, which include acetylcholine, noradrenaline, serotonin, dopamine, and, most likely, GABA, are produced by clusters of neurons located in a variety of subcortical structures. By means of their widespread axonal projections, these neurons deliver the modulator neurotransmitters throughout the brain. Most studies have so far focused on the role of these chemicals in determining a particular state of sleep or wakefulness, but I believe that it is the *collective* action of these modulatory structures that triggers the shift from one brain state to another, leading to an equivalent change in animal behavior.

In this way, the Loch Ness state-space is a condensed depiction of global brain dynamics and how these dynamics are determined by the energy available to the brain. From moment to moment, a subject's global brain dynamics varies as it responds to the neuromodulatory influences that "push" the generation of continuous electrical activity among billions of interconnected neurons. The brain, however, can only move from one stable dynamic state to another equally stable state, assuming, that is, that the cortical circuits can together reach the necessary energy threshold to do so. This happens when the forebrain as a whole attains a high level of coherent and synchronous neuronal activity.

By delivering this level of descriptive detail, it's no wonder that the Loch Ness monster quickly became our favorite lab pet.

Although the state-space may sound abstract, it opened the way to understanding what happens when identical physical stimuli from the outside world arrive to the body periphery during different states of global brain dynamics. In other words, the state-space plot gave us a window into the brain's own point of view.

We have already seen how information from the outside world is detected and transduced by the body's peripheral sensory organs into a spatiotemporal stream of incoming electrical activity that ascends through the sensory pathways to the brain (see chapter 5). Naturally, when this incoming spatiotemporal signal hits the cortical circuitry, the brain is settled into a particular dynamic state. Following the relativistic brain hypothesis, it is the collision of these two spatiotemporal signals, the incoming peripheral signal and the internal dynamic state of the brain, at a given moment in time that generates the actual pattern of electrical activity that morphs into one's perception of the world. Accordingly, we would expect that identical ascending peripheral sensory stimuli reaching the cortex at two distinct internal dynamic states should produce totally different patterns of activity, and therefore induce different perceptual experiences in the same subject.

As we saw in chapter 5, even before we had discovered a way to recognize the Loch Ness monster in our state-space plots, we passively delivered identical tactile stimuli to the infraorbital nerve, the branch of the trigeminal nerve that innervates the rat's facial whiskers, during the quiet-awake, active-exploration, and whisker-twitching states. When these identical stimuli hit the brain at these distinct states, they evoked very different sensory neuronal responses, at both the cortical and thalamic levels. In a follow-up experiment, we looked at whether it made a difference that the stimuli had been delivered passively. We again observed that both cortical and thalamic neuronal tactile responses differ dramatically—particularly when the animal has actively engaged the tactile stimulus through whisking. In a further experiment, conducted with a neighboring laboratory led by my best buddy, Sidney Simon, at Duke's Department of Neurobiology, Jennifer Stapleton demonstrated that the physiological properties of the sensory responses of single neurons in the gustatory cortex also differ dramatically, depending on whether an animal is actively licking a tube to sample a tastant or simply receiving a sample of the same chemical passively inside its mouth.

Our findings have been corroborated by examinations of brain activity during slow-wave sleep, when the thalamocortical loop becomes entrained by highly coherent synchronous sleep spindles. This results in an almost complete interruption of the sensory volleys coming into the cortex. Indeed, some neuroscientists have classified this phenomenon as a "functional disconnection" between the peripheral and the central nervous system, since no ascending sensory signal seems able to cross the thalamic level and reach the primary sensory cortices. Similar findings have been reported for both the visual and auditory systems, where the flow of incoming sensory stimuli is affected by the animal's brain state.

However, global brain dynamics define only one component of the brain's own point of view. Embedded in these internal dynamics are a heap of memories, accumulated throughout the animal's previous life experience. This mnemonic information also contributes to the spatiotemporal collision of incoming peripheral signal and internal dynamic states. According to the relativistic brain hypothesis, the animal's mnemonic existence is felt even before the incoming sensory signal hits the forebrain, by dictating the generation of an anticipatory signal across the cortex and likely most of the forebrain. This electrical distortion, which likely includes components of the spatiotemporal motor signals created by the animal as it explores its environment, may account for the fact that neuronal activity is modulated across all layers of the rat S1 cortex, as well as the somatosensory thalamic nuclei, several hundred milliseconds prior to the moment when the animal touches any object with its whiskers. Under the influence of this "expectation" signal, which can either increase or decrease the firing rate of single neurons across a population, the internal brain state is pre-adjusted, creating the brain's initial model of the external world. In this sense, the relativistic brain "sees" before it "watches," and the brain exerts its own point of view.

I believe that the matches and mismatches between these two spatiotemporal signals, one generated inside the brain and the other from the outside world, ultimately define what we perceive as reality. That implies that there is no absolute truth, because the brain is not a mere slave to what, for example, our retinas report to have seen. This neurophysiological collision is encapsulated in the *context principle.*

THE CONTEXT PRINCIPLE

The way the cortex responds as a whole, to an incoming stimulus or the need to produce a particular motor behavior, depends on the global internal state of the brain at that instant; that is, ongoing brain dynamics are essential in defining the optimal solution the brain derives for the generation of any given behavior.

I should emphasize that the Loch Ness state-space, in all likelihood, reveals just a few of the internal or hidden dynamic states of a brain that supply the context for thinking. In reality, there must be tens of other states embedded in our plot that cannot yet be as readily distinguished. For example, smaller, more fine-tuned dynamic brain states likely reside within the large crowd of dots that defines the time during which animals are awake. That remains a question for further experimentation and analysis.

At this point it may be useful to introduce a graphic model I have been using for the past few years to visualize these spatiotemporal collisions, the energy limitations on the brain, and, above all, my metaphor of the brain as a sea of electricity bound by waves of neural time. I call this graphic representation the "wire and ball model" (see Fig. 11.4).

In the wire and ball model, a closed wire loop ascends, dives, and twists all over a three-dimensional volume, tracing the viable pathways that the brain can take within the Loch Ness state-space to move from

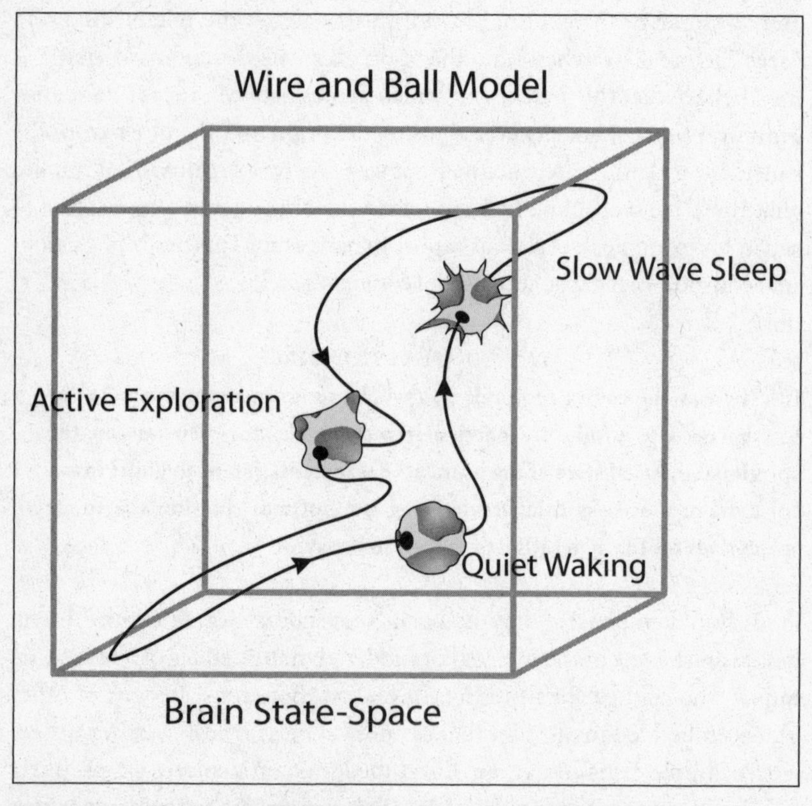

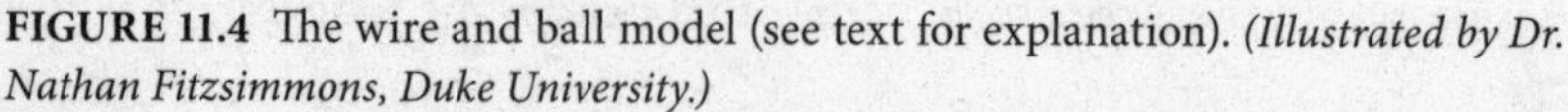
FIGURE 11.4 The wire and ball model (see text for explanation). *(Illustrated by Dr. Nathan Fitzsimmons, Duke University.)*

one internal brain dynamic state to another. As such, the bending and twisting of this closed-loop wire can be thought of as the track of a dynamic roller-coaster of electrical activity that the brain rides during its wake-sleep cycle.

Now, let's add a ball to this model. Conveniently, the ball has a hole in it, so it can slide freely along the wire, wherever the wire will take it on its neuronal roller coaster ride. The ball represents a large random sample of cortical neurons, a good chunk of the brain, if you will, that belongs to a particular brain region, such as the primary somatosensory cortex. By definition, the sphere's total volume is kept constant as it slides along the wire loop. That's because the ball represents the maximum number of spikes that this sampled neuronal mass can produce, which is capped by the amount of energy available to the brain. The most interesting property of the ball, however, is that its overall shape varies as a function of the pattern of electrical activity generated by this particular neuronal population. Thus, as the ball rides along the wire, moving from different internal brain states by climbing or descending through the transitions that flank them, its three-dimensional shape will change considerably, though its overall volume will remain constant. Thus, when the ball is within the boundaries of the quiet-awake state, characterized by low-amplitude, high-frequency neuronal oscillations, the ball's surface is relatively smooth, much like the surface of the sea on a peaceful, windless day; just small, high-frequency ripples appear. In contrast, when the ball moves into the slow-wave-sleep state, it must first cross through a high-coherence transition, with high-amplitude spindle waves riding on top of low-frequency delta oscillations. The sphere's surface in this case would be distorted, with crisp, spiky crests spreading across it while exposing the troughs within the ocean currents. That is because as we track the ball's trajectory through the wire loop, the three-dimensional representation of the spatiotemporal activity of the neuronal population is constantly influenced by the changes in internal brain dynamics that define the state-space.

Now, suppose we add a fourth dimension to the ball. This is done by creating a color scale that measures how fast each neuron of the sample is firing in a fixed period, say one second, around the time an incoming spatiotemporal stimulus collides with the ball as it travels along its never-ending wire roller coaster. In this color-coded scheme, dark gray represents a very high firing rate compared to average, and light gray represents a very low firing rate, with intermediary shades assigned between these two extremes. The gray shades of this imaginary, and admittedly very strange, ball change depending on where it is on the wire when a sensory stimulus from the outside world happens to hit it. If the animal is in the quiet-awake state—alert, but not actively exploring the world—the arrival of the incoming stimulus will cause the individual S1 neurons represented by the ball to quickly increase their firing rates. This firing will be sustained for a very brief period of time, before the rate rapidly falls below its original level, characterizing a period of post-excitatory inhibition that lasts for tens of milliseconds. The wire and ball model is up to the challenge of depicting this well-documented sequence of events. First, a dark gray bulge suddenly emerges from the surface of our sphere, representing the initial intense firing response to the incoming stimulus. This deformation is rapidly followed by the appearance of a light gray sinking hole. The more individual neurons that are recruited by the incoming stimulus, the larger the dark gray bulge and the light gray sink.

Things would look very different, however, if an identical incoming spatiotemporal tactile stimulus hit the same ball of cortical tissue while it rides the region of the wire analogous to active exploration. During this brain state, neurons modulate their firing rate up and down long before an incoming tactile stimulus hits the S1 cortex. For that reason, our ball's surface would start bulging and sinking in advance of its collision with the stimulus. By the time of the collision, the ball would be deformed, in a complex and rather colorful manner, so that it barely resembled the ball as it looked when a stimulus occurs during the quiet-awake state. This would especially be true because different numbers of neurons would be recruited to respond to the same whisker stimulus depending on whether the rat was in the quiet-awake or active-exploration state.

As you can see, everything is relative, when it comes to the processing of incoming sensory signals by the internal dynamic state of the brain. For instance, Figure 11.5 depicts samples of a single neuron's firing at one-second time periods. The top trace shows a sample of the firing of this neuron during the quiet-awake state. As with any other neuron, this cell has a limit on how fast it can fire, defined by its refractory period, the period that follows the last spike produced by the neuron and which is used by the neuron to recharge its membrane capacitor. Assuming this neuron's refractory period is two milliseconds, its maximum instantaneous firing rate (or spike velocity) is limited to five hundred spikes per second. Notice how, in the top portion of this figure, the individual action potentials produced by the neuron during the quiet-awake state are spread out over the entire second we analyzed, and that most of the time intervals between consecutive spikes are very long, usually involving tens of milliseconds. This neuronal spontaneous firing pattern approximates a Poisson process, in which a sequence of events occurs continuously but independently.

Now imagine that at the exact end of the first one-second period a strong tactile stimulus collides with the internal dynamic state of our chosen neuron. The neuron responds robustly to the colliding tactile stimulus, with several action potentials generated early in the second firing we recorded, as seen in the middle portion of the figure. Instead of being dispersed over the whole second of time, as one might expect from the neuron's previous activity, the distribution of spikes shifts dramatically, so that many of the spikes are crowded around the onset of the stimulus. This means that each consecutive action potential in this early portion of the time segment is separated from the next by just a couple of milliseconds. The neuron is firing at a much faster velocity, based on its own internal time reference. This appears to continue until the neuron reaches its maximum instantaneous firing velocity of 500 spikes per second, a biophysical wall based on the limits of the neuron and the overall energy available to the brain.

After this initial salvo or burst of action potentials, there is a long period, lasting up to one hundred milliseconds, in which no spikes are produced by the neuron. According to the relativistic brain hypothesis, this silent period is imperative: since the firing is bounded by a cap in the total number of spikes a neuron can produce—akin to a "neuron spike budget"—and a maximum instantaneous spiking velocity, to take part effectively in the representation of an incoming sensory signal the neuron has to "borrow" a few action potentials from its total spiking

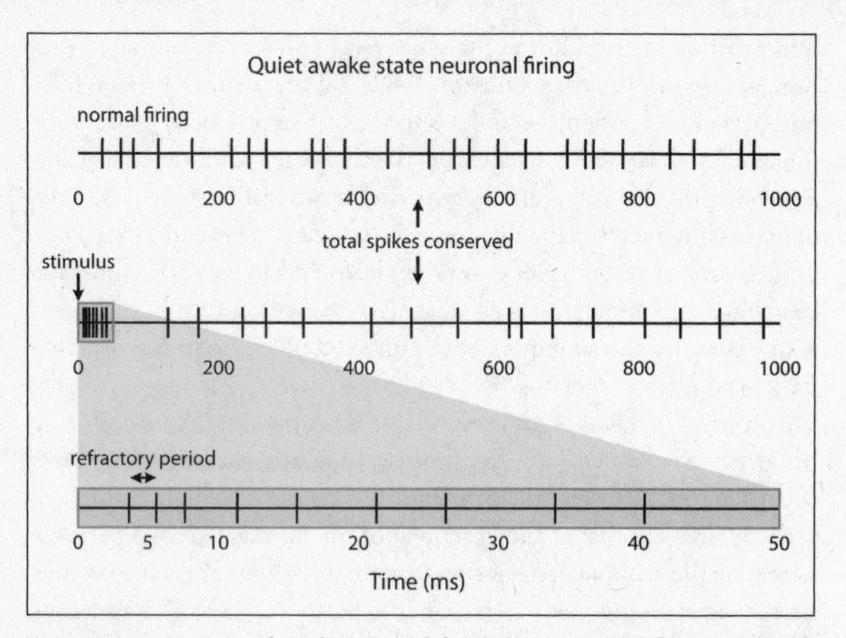

FIGURE 11.5 Patterns of single neuronal firing during normal firing (top shelf) and during the occurrence of a sensory stimulus (middle shelf). The refractory period, illustrated in the bottom shelf, defines the maximum firing rate that a neuron can reach. Throughout these examples, the total firing rate produced by the whole brain has to be maintained below a maximum cap. *(Illustrated by Dr. Nathan Fitzsimmons, Duke University.)*

budget. Normally, the neuron spike budget would be spread across a somewhat long period of time—one second, as in the example above. Now, however, these potentials need to be delivered in a hurry, effectively warping the neuron's internal timing reference.

Finally, the wire and ball model can illustrate exactly what happens when a tactile stimulus is delivered while an animal is maintained under deep anesthesia. When oscillations are recorded from anesthetized animals, the entire Loch Ness state-space all but collapses, transforming it into a single, impoverished mental blob. Since different anesthetics affect neuronal pathways in a variety of ways, when a subject is anesthetized the neurophysiological measurements of the receptive fields of single neurons and the resulting sensory maps are severely circumscribed and sculptured. This results in a highly localized representation of information, wherein the dynamic firing patterns of populations of neurons are not captured because those neurons have been deprived of their means to interact. Far from being dynamic, this impoverished state renders the experimental subject unconscious and unresponsive, and robs the brain of its mind.

When these firing constraints and behaviors are applied to all of the single neurons interacting in a neural ensemble, two more interrelated neurophysiological principles of the relativistic brain are established: the *conservation of neural ensemble firing principle*, which has been verified in multiple species and multiple cortical areas in our laboratory, and the *neuronal mass effect principle*, derived from our analysis of the neuronal dropping curves calculated during operation of BMIs.

THE CONSERVATION OF NEURAL ENSEMBLE FIRING PRINCIPLE
Not only is there a maximum limit of firing that an ensemble of neurons can reach, but the global ensemble firing rate tends to stay constant, hovering around a mean, due to a variety of compensatory mechanisms that create a stable equilibrium. If single or multiple cortical neurons increase their firing rate instantaneously, an equivalent mirror image reduction in firing is soon produced by other members of the neural ensemble so that the overall energy budget of the brain stays constant over the long run.

THE NEURONAL MASS EFFECT PRINCIPLE
As the size of cortical neural ensembles grows beyond a certain large number, the amount of information embedded in the neural ensemble tends to asymptote, slowly converging to its maximum information capacity. This effect is reflected by an important reduction in the statistical variance of prediction derived from large neuronal ensembles. The neuronal mass effect principle is one potential way to explain how the considerably higher variance of a single neuron's firing ends up being washed away whenever that individual neuron's contribution is averaged across a large neural ensemble to which the single neuron becomes functionally attached during the execution of a particular behavior.

To confirm the conservation of neural ensemble firing in the human cortex, neuroscientists will need to turn to functional magnetic resonance imaging (fMRI), a technique used to measure blood flow—that is, brain metabolism and energy consumption—in real time. My prediction is that for each cortical spot found to have increased its metabolic activity, there should exist an equivalent mass of neurons that show a decreased metabolism.

Given the constraints imposed by fixed spike budgets and fixed maximum neuronal spike velocities, the shape of the distribution of spike timing has to be adjusted to allow a neuron to represent a tiny fraction of the information contained in the stimulus. Moreover, the spatial recruitment of neurons that signals a particular collision of an incoming stimulus with the ongoing activity of a particular internal dynamic brain state varies from moment to moment. Indeed, in a series of experiments we have seen that if identical tactile stimuli are delivered at different internal brain states, the overall spatial size of the cortical population recruited to participate in the representation of this stimulus varies. In addition to the overall size of the population, the particular spatial distribution of firing within this population also changes from moment to moment. The relativistic brain hypothesis suggests that a great deal of this variability is dictated by a delicate energy balance. Thus, if a particular neuron "borrows" a lot of spikes from its own budget to produce an exuberant firing response to a stimulus, another neuron needs to reduce its spiking rate to "finance" the firing of its neighbor. To some degree this sounds like how the global financial system works today: while a few bankers suck all the money from the market as if they had some sort of nuclear-powered vacuum cleaner, millions find themselves, all of a sudden, without any money whatsoever.

This idea owes a great deal to one of the classical concepts in systems physiology, the *milieu interieur* (later known as body homeostasis), developed by the incomparable French physiologist Claude Bernard. An almost forgotten commodity in the reductionist-obsessed medical school curriculum, homeostasis plays a key role in regulating how our dynamic body states are managed and maintained at each moment of our entire conscious life. Effectively, this means that mechanisms that maintain internal brain homeostasis, particularly energy consumption, may dictate the limits of complex information processing by the brain. I propose that in a relativistic brain the information generated by a subset of billions of neurons is a function of the amount of energy these neurons collectively consume at each moment in time.

Finding the mathematical relationship between these two quantities, if it actually exists, would be a major breakthrough in modern neuroscience. But would it unlock the mysteries of the egotistical human brain?

Only time will tell.

12

COMPUTING WITH A RELATIVISTIC BRAIN

On June 21, 1970, in the midst of a heat wave that insisted on sucking what little oxygen remained at the high altitude of Mexico City, the Brazilian and Italian national soccer teams played what is considered by many as the most memorable championship game in the history of the World Cup. Although the game was going to be played on just another sultry summer afternoon at Azteca Stadium, a morning shower had made the pitch slick and somewhat unpredictable. Both the Brazilians, led by Pelé, and the Italians, captained by the legendary leftback Giacinto Facchetti of Internazionale de Milano, were playing for their third title and a right to keep the famous Jules Rimet trophy for good.

After a hard-fought first half that ended in a 1–1 tie, the Brazilian team, wearing their traditional canary yellow jerseys stamped with green numbers, pulled off a formidable showing in the second half. With the game clock counting down the last four minutes of regular play, the score stood at 3–1 Brazil. The win was all but guaranteed.

But winning had never been enough for the Brazilians. That may explain why their last big play of the match seemed to unfold almost like magic, when Tostão sneakily robbed the ball from a tired Italian forward, next to the left edge of the Brazilian penalty box. Having gained possession, he flicked an inconsequential pass to Piazza, who passed the ball, without much fuss, to Clodoaldo, who himself promptly passed the ball to Pelé, deep in Brazil's half field. Upon making contact with his lifelong companion, Pelé employed one simple foot caress to deliver the ball to Gérson, who hastily transferred it on to Clodoaldo. Soon enough, it became evident that the usually cerebral Gérson had created an ill-posed problem for Clodoaldo's motor cortex. Surrounded by four Italian players, all desperate to strip away the ball, Clodoaldo did not flinch. Upon securing the ball with his reliable right foot, he started dribbling his way out of that crowd of rivals, one by one. For the next five seconds, he danced with the ball until the last of his opponents retreated in despair.

Recovering his balance and cool, Clodoaldo spotted Roberto Rivelino, free on the left wing. Unselfishly, he passed the ball to *La patada atomica* (literally, the atomic kick), a nickname received in honor of Rivelino's prodigious and lethal left shooting foot. Despite these accolades, Rivelino looked tired when he received the ball. Mercifully, he did not need to hold it for long. Jairzinho—*Furacao* (the hurricane)—was waiting, with plenty of energy to spare, in the left flank of the Italian defense. Without his looking at the ball, Jairzinho's brain assimilated the tool of his trade with a quick sequence of touches, then daringly made a move to his right, where, unexpectedly, a defensive midfielder was stalking. Jairzinho had no choice but to resort to an ugly, childhood trick, the "toe poke," to pass the ball straight to the immortal right foot of Pelé, who had run the extension of the pitch to position himself immediately in front of the Italian penalty box, pretty much like a jaguar ready to feast on its prey.

Pausing to caress the ball with his foot, Pelé barely noticed when Tostão, fending off a defender, screamed and pointed anxiously at an opening to score. Pelé acknowledged Tostão's warning by redirecting the ball, with almost effortless disdain, to a vast open area of field to his right. Just as those watching started to think the king of football had gone mad, the ball reached the empty spot that Pelé's brain had plotted out for it. There, it was instantaneously met in midair by the charging right foot of Carlos Alberto, the Brazilian team captain, who had played with Pelé for the best part of ten years. What came next was the only thing predictable in that exuberant display of collective play: Goooaaalll!

As the announcer's crescendo echoed, an entire country plunged into the streets, dancing.

That was quite a goal. Thirty seconds of uninterrupted play. Eight different players touching the ball. None having the slightest idea, beforehand, what the outcome of their collective interaction would be. That unforgettable sequence of movements, and the 4–1 score it sealed, demonstrate the insurmountable power unleashed by the emergent behavior of a dynamic complex system. No matter how proficient and intelligent each of those players were individually, the play they engendered together could never be predicted or planned a priori—it simply emerged from their combined voluntary actions, in hundreds of milliseconds, before they even realized it into a conscious thought.

Faced with the unpredictable behavior of such a complex system, the classical reductionist strategy would break it down into its smallest free-standing components and then, after fully characterizing the properties of these elements, try to derive the success of the entire system from the principles underlying its individual parts. That approach would completely miss how the Brazilians made their play and won their third World Cup that afternoon. For example, let's imagine, rather logically, that each player is the smallest individual component comprising the Brazilian team. Reductionists would prescribe that, in order to understand the mechanisms that generated that play, one should collect all of the data available for each of the players involved in the goal, from their physiological descriptors to their individual passing, shooting, and scoring records. Next, one could try to describe each player's average reaction time, muscle metabolism, and motor control capabilities. The players' behavior in previous championship games would be studied, to characterize their decision-making processes. A particularly enthusiastic devotee might suggest that a complete answer would only be possible by analyzing each player's entire genome. Pretty soon, mountains of information about each player would be assembled. As the British physicist John D. Barrow writes in his delightful book, *Impossibility*, "All the examples we have listed are made of atoms if you look at a low enough level, but that does not help us to understand the distinction between a book and a brain."

Despite all this effort, no one feeding the accumulated data into a computer would be able to predict how a team made up of those players would have behaved in a soccer match. The behavior of the Brazilian team would be labeled as yet another of the noncomputable phenomena of nature. No one would be able to reconstruct or predict that dramatic example of team play by using a reductionist strategy because the play emerged from the unpredictable, dynamic interaction of eight interconnected variables, who happened to be the best soccer players in the world in the year 1970.

Take, for instance, the huge number of potential plays that can be generated by the interactions of the eleven positions of a regulation soccer team—a number so large that it, too, is incalculable. In his book, Barrow describes the enormous degree of complexity that can be generated by a system formed by a large number of connected elements. "Complex structures seem to display thresholds of complexity which, when crossed, give rise to sudden jumps in the complexity," he argues. "One person can do many things; add another person and a further relationship becomes possible; but gradually add a few more people and the number of complex interrelationships grows enormously. Economic systems, traffic systems, computer networks: all exhibit sudden jumps in their properties as the number of links between their constituent parts grows." And as complexity jumps, so does the unpredictability and the information content of the system. Such jumps are part of the history of the universe. In the beginning there was only physics. Then chemistry appeared. Biology evolved next and when consciousness emerged, everything in our part of the cosmos changed dramatically. At least from our own humble perspective.

If it seems so difficult to predict the collective behavior of some subset of eleven people, how could it ever be done for a dynamic, ever-shifting population of neurons drawn from the tens of billions in the human brain? This is the dilemma that systems neuroscientists have confronted for most of the past century. By focusing on single neurons, neuroscientists may have gathered an impressive wealth of information about the biological properties of the individual processing units of brain circuits—a rewarding and very useful endeavor. Perhaps looking at the operation of single neurons was the only safe and technically feasible path to move forward, for a long time. But, as Barrow appropriately states, "consciousness is the most spectacular property to emerge" out of a complex system. By turning the gaze of ever more powerful tools onto the single neuron alone, neuroscience relinquished, almost without knowing, any real concrete chance of understanding the physiological mechanisms of thinking, the main product of those vast neuronal galaxies that define the conscious inner universe that exists within our heads.

As we have seen over the course of this book, in the past two decades my lab has been part of a new wave of neuroscience that is changing the way we approach the brain, using new measurements of complex neural ensemble behavior to trace the interwoven sources of thinking, including all of the interdependent, relativistic dynamics that change, from moment to moment, the way in which neuronal space and time fuse, bending and shifting the contribution of each component of the circuits that define the nervous system. To return to my soccer analogy, while no one could predict how that particular play at Azteca Stadium emerged from the collective work of the Brazilian team, we certainly can say a few things about the particular conditions in which the play took place. For instance, the players had an ultimate, shared goal: to score more goals than their rivals, win the game, and carry the World Cup home. Moreover, the team was formed by very experienced players, some of whom had played hundreds of games together, "gelling"—an expression used by soccer fans to describe a team that plays well together. These Brazilian artists had learned, over their careers, how to select the kind of team strategy that would work (or not) under the conditions of that pivotal match. Having watched a few of the Italians' previous games during earlier World Cup rounds, the Brazilians also shared knowledge of their opponents' usual defensive and offensive tactics. Thus, prior to their encounter with the Italians, the Brazilian team had built an elaborate array of expectations, a "mental game model," of what the other team might muster in a variety of situations. Now, facing only four minutes left in play, and seeing that the Italians were visibly tired, the Brazilians adjusted their game model and selected an optimal "last-dash" strategy to seal their victory. Their magnificent play was influenced by a common, overarching goal; a foundation of previously built expectations; a game model shaped by physical potential and constraints as well as adaptability; and the collective ability to interpret the particular context encountered at a particular moment. As any soccer "scholar" would attest, the context presented to the Brazilians dictated as the optimal strategy that the team spread itself across the pitch and use long but highly precise ball passes to "stretch" the Italian defenders as well. That ensured that each tired Italian had to run a lot farther for the mere opportunity to steal the ball from a Brazilian. Once you take into account all these factors, the number of viable plays that could have emerged from the masterful Brazilian team is significantly reduced—not enough to predict the precise play, but enough to suggest that some plays were more likely to happen than others.

Throughout this book, I have defended the position that the brain's own point of view decisively influences the way each of us constructs a model of reality. Like the dynamic, interconnected players of the 1970 Brazilian soccer team, I believe the brain achieves its goal through the emergent properties of a highly complex system.

In soccer, the players interact on the field of play, following an infrastructure (the rules of the game) and pushing against physical constraints (from the power produced by their bodies, to their maximal speed across the field, to the force of gravity). They are presented with a fielding or scoring opportunity, and produce an optimal solution based on their game models. Throughout, the system remains relativistic, with actions emerging from collective thoughts in the continuously shifting context of space and time.

I propose that the brain similarly achieves its point of view, working within its own operational constraints and physiological infrastructure to create a variety of behaviors out of the complexity of the nervous system. So far I've introduced ten principles that go a long way, in my opinion, to describing how thought comes alive, in the truest, electrical sense. These are somewhat like the rules of the game, but underlying them all are two simple anatomical and physiological facts that, like gravity and electromagnetism in the larger natural world, define the organic universe in which the brain operates:

1. Billions of neurons produce electrical currents that are capable of spreading through the continuous, salty, and hence highly conductive space between and around the brain's densely packed cells, generating widespread electromagnetic fields that, despite being very tiny in absolute magnitude, can still influence neighboring neurons.
2. A hugely complex network of tens of thousands of potential long-range feedforward and feedback connections, which include multiple multisynaptic cortical and subcortical loops, provides thousands or even millions of ways through which neurons in a given cortical area can readily communicate with other neurons located at a relatively great distance, far away in the brain (a fact I discovered when my thesis research produced those cascades of printouts of possible neuron-to-neuron connections).

These two basic elements help to explain, for instance, findings like those that shocked Eberhard Fetz and his colleagues in 1998. In their study, Fetz's group learned they could recover information about a visual cue that had been used to train a monkey to make an arm movement. That may not sound at all surprising, given the experiments I have recounted—except that they were recording the activity of interneurons located in the intermediary layers of the spinal cord. Before these results, no neurophysiologist would have claimed that the spinal cord had anything to do with vision, yet visual information could be recovered from cells there. Furthermore, these same general properties of brain connectivity may explain why two postdoctoral fellows in my lab, Romulo Fuentes and Per Petersson, and I discovered that stimulation of the dorsal surface of the spinal cord can dramatically reduce the tremors, akinesia, and other symptoms of a neurological syndrome, similar to Parkinson's, that affects mice and rats depleted of dopamine. Stimulating the spinal cord appears to produce a powerful disruption in the patterns of epileptic-like electrical activity crossing the animals' motor cortex and the basal ganglia. In the relativistic brain, it seems, many roads take you from nowhere to everywhere else.

More proof that the brain does not respect the borders created by cortical localizationists comes from repeated independent observations of cross-modal cortical processing in primary sensory fields—in stark contrast to classical hierarchical doctrine, which states that cross-modal processing should only take place in so-called higher-order associative areas in the cortex. In the mid-1990s, instances of cross-modal processing in the visual cortex began to be reported among patients suffering from definitive visual deficits (for instance, congenital or acquired blindness) or people submitted to temporal visual deprivation during experiments. In one study, published in 1996, Norihiro Sadato, Alvaro Pascual-Leone, and others working in Mark Hallett's group at the National Institute of Neurological Disorders and Stroke (NINDS) employed a brain-imaging technique known as positron emission tomography (PET) to demonstrate that both the primary and secondary visual cortex were strongly activated in people who, after becoming blind in early life, had become proficient Braille readers, when they performed tasks that required fine tactile discrimination. A year later, the same NINDS research team decided to "disrupt" activity in the V1 cortex while a blind person was presented with Braille letters or embossed Roman numbers to read. This disruption was induced by a technique called transcranial magnetic stimulation (TMS), which, as the name implies, delivers magnetic pulses, in a noninvasive way, to a specific cortical region, causing an interference in normal neuronal activity. In subjects with no visual impairment, TMS targeted at the V1 cortex led to problems with the visual recognition of letters, but had no effect on the ability to discriminate between different tactile information. When the researchers presented blind individuals with the analogous task while the V1 cortex was disrupted, however, the subjects committed significant discrimination errors—even though the task involved tactile information. This suggested that the enhanced ability displayed by blind individuals in tactile discrimination tasks, such as Braille reading, may emerge because the visual cortex is recruited to help out—what is called *cross-modal recruitment*.

In another study, David C. Somers and his colleagues at Boston University showed that after a mere ninety minutes using a blindfold, the primary visual cortex of otherwise normal-seeing human subjects became activated when subjects performed tactile tasks. Such a short period of deprivation, insufficient for a change in gene expression or the creation of new anatomical connections, favored the idea that cross-modal responses were based on previously existing somatosensory afferents in the visual cortex. Blindfolding plainly, if ironically, unmasked the ability of the V1 cortex to process tactile information.

More recently, Sidarta Ribeiro's group at the Edmond and Lily Safra International Institute of Neurosciences of Natal, Brazil, has extended these findings by demonstrating that cross-modal responses are present in both the primary visual and primary somatosensory cortices of intact rats—meaning rats without any transient or permanent sensory deprivation (see Fig. 12.1). Ribeiro identified individual neurons in the S1 cortex that readily respond to visual stimuli and single neurons in V1 that readily respond to the stimulation of a rat's whiskers. Ribeiro observed cross-modal neuronal responses like these even under anesthesia. As these rats awake and start performing different tactile discrimination tasks in the dark, large numbers of individual neurons in V1 continue to fire in response to pure tactile stimulation of the animal's facial whiskers. More extraordinarily, when an intact rat uses its whiskers to discern the size of an opening diameter in total darkness, the cross-modal tactile responses across V1 neurons contained enough information to predict the opening's diameter, and the performance of these V1 ensembles was just as good as that of an equivalent population of S1 neurons.

Ribeiro's results have been bolstered by experiments conducted by Yong-Di Zhou and Joaquín Fuster at the University of California, Los Angeles, who reported that visuotactile cross-modal associations develop in the primary somatosensory cortex of intact rhesus monkeys as these animals are trained to perform tasks that require associations involving both vision and touch. And it appears these cross-modal responses in S1 can be enhanced by training animals in tasks whose contingencies, by design, include cross-sensory associations.

Such a view is also supported by studies involving the primary auditory cortex (A1). In a beautiful series of experiments in rhesus monkeys, Asif Ghazanfar has shown that multisensory integration of dynamic facial attributes and voices in A1 may contribute to the way primates, including humans, communicate with one another.

Even the neurons of the primary gustatory cortex respond to a vast repertoire of multimodal sensory responses.

All of this is a far cry from the cytoarchitectonic divisions of the

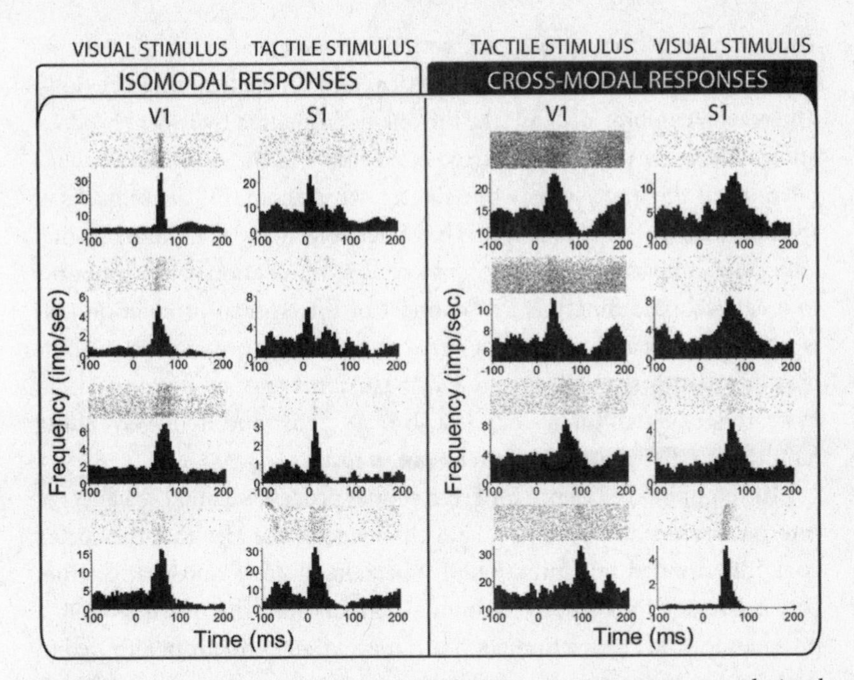

FIGURE 12.1 Cross-modal processing in the primary somatosensory and visual cortices of rats. Peri-event histograms display the isomodal and cross-modal sensory-evoked responses of individual S1 and V1 neurons. In the left panel, the traditional visually evoked responses of V1 neurons and tactile-evoked responses of S1 neurons are shown. In the right panel, samples of V1 neurons that respond to a tactile stimulus and S1 neurons that respond to a visual stimulus are plotted. Korbinian Brodmann would be shocked! *(Courtesy of Dr. Sidarta Robeiro, International Institute of Neuroscience of Natal, Brazil; and Dr. Damien Gervasoni, Claude Bernard University, Lyon, France.)*

brain perpetuated by Brodmann and generations of localizationists, who have claimed that rigid anatomical and functional borders exist within the cortex. Such a functional model of the brain, devoid of time and internal dynamic brain states, served us well for a century. Lately it has become a major hindrance for progress in our thinking about how the cortex processes information in natural ethological conditions.

At this crucial juncture you may be asking yourself, "But what about Paul Broca's patient in the nineteenth century who could not speak anymore following a lesion of the left frontal lobe? Doesn't that finding still support strongly the localizationist point of view?" Actually no.

Today we know that speech production depends heavily on the concurrent interaction of a multitude of cortical and subcortical brain regions. The reason cerebral strokes, like the one documented by Broca, produce aphasia is likely that they destroy, in addition to the gray matter, huge portions of the underlying white matter, which contains dense packs of the nerve fibers that connect this huge network of areas with the frontal lobe. Such a massive destruction of key communication cables amounts to a catastrophic functional disconnect of this speech production network. Although Broca's patient survived this rather massive stroke, he may have lost his speech through a catastrophic cortical disconnection. With this explanation, at long last, Broca's ghost and its never-ending haunting of distributionists can be put to rest.

The combined evidence for cross-modal processing and the effects of internal brain states mounts a fatal challenge to the idea that the cortex is rigidly divided into functionally specialized areas and that distinct cortical regions are purely unimodal. Other than the neatness that it provides for the localizationists' brain maps, there's no profit afforded to such a unimodal approach, simply because there is no experience in our lives that can be defined as purely unimodal. Moreover, the real world is not made of a conflagration of circular flashes or rectangular bars of light, localized skin indentations, pure auditory tones, or primitive tastes and odors, and most of the time we live our lives as freely behaving beings, not in a deeply anesthetized state. Only in neurophysiology laboratories have these conditions been created, and more and more I believe that these artificial worlds have meant that we have been studying a completely different kind of brain—certainly not the one that all of us depend on to survive day in and day out.

I propose that the brain we do use—the relativistic brain—is more akin to a medium in which neuronal space and time fuse into a *physiological space-time continuum*, which can be recruited in a variety of ways to perform all the tasks assigned to it. Depending on the status of the peripheral sensory organs, the task demands, and the brain-state context in which behaviors have to be produced, this physiological space-time manifold can be dynamically twisted, bent, and shaped in an optimal information processing configuration that, at any given moment, endows us with our best neuronal shot to achieve our goal-oriented behaviors. This notion of a cortical neuronal space-time continuum is totally compatible with the existence of ripples of probabilistic regional functional specialization. Yet, in this new conception such ripples are neither absolute nor immutable for the duration of one's life. Instead, they can shift quickly, according to the task at hand.

THE NEURONAL SPACE-TIME CONTINUUM HYPOTHESIS

From a physiological point of view, and in direct contrast to the classical twentieth-century canon of cortical neuroanatomy, there are no absolute or fixed spatial borders between cortical areas that dictate or constrain the functional operation of the cortex as a whole. Instead, the cortex should be embraced as a formidable, but finite, neuronal space-time continuum. In this continuum, functions and behaviors are allocated or produced respectively by recruited chunks of neuronal space-time, according to a series of constraints, among which are the evolutionary history of the species, the layout of the brain determined by genetics and early development, the state of the sensory periphery, the state of the internal brain dynamics, other body constraints, task context, the total amount of energy available to the brain, and the maximum speed of neuronal firing.

Essentially, the cortex should cease to be treated as a hierarchical mosaic of discrete, segregated, highly specialized, and virtually autonomous cortical areas.

Unlike previous incarnations—notably Karl Lashley's equipotentiality theory—my concept of a neuronal space-time continuum has no qualms in accepting that there is some degree of cortical specialization, dictated mainly by the general strategy through which the cortex and thalamocortical connections were laid down during early postnatal development. But development is not destiny, and populations of neurons can be recruited as needed once the initial cortical layout is set down. Those ontogenetic specializations, like a featured soloist, sit atop a powerful symphony of multimodal and dynamic cortical interactions that dictate how a relativistic brain works throughout its unique existence.

Recently, during one of my talks, a very distinguished cognitive neuroscientist, who seemed befuddled by my notion of a neuronal space-time continuum, asked me to explain one nagging paradox he had detected. Why would nature invest so much energy during early development to building all these highly segregated sensory pathways, not to mention topographically organized cortical maps, and then, out of nowhere, decide to relinquish these efforts for the relativistic mess of brain dynamics I proposed? In response I volunteered that, as far as I could say after twenty-five years of seeing, listening, and recording brainstorms, waves of cortical spikes do not appear to stop at, or care about, the aesthetically pleasing borders of old-fashioned cytoarchitectonics. Instead, they simply pass through them, as if those borders were mere fantasies created in someone else's brain.

Assuming that the relativistic brain hypothesis and the neuronal space-time continuum deserve further exploration, I have dedicated the next and final chapter to speculate on what, if anything, could emerge from the interfacing of a relativistic brain with the most elaborate and smart computer ever dreamt (but not yet built) by humanity. But before we get there, I would like to discuss why I believe *relativistic* is the best word to describe the way our primate brains operate.

In an engaging analysis of the "main philosophical impulses" of relativism, the Irish philosopher Maria Baghramian lists three attributes, among many others, that are pertinent to a relativistic view of the brain: context-dependence, mind-dependence, and perspectivalism. *Context-dependence* refers to the fact that many (if not most) human decisions and judgments, as well as the expression of one's dearest beliefs, are influenced by "events that happen at a particular time and place and to a particular person." *Mind-dependence* involves the long history of philosophical thought that espouses the position that the human view of reality, as well as our judgments, beliefs, explanations, and scientific theories, is irrevocably tainted by the heavy touch of bias imposed by the human mind, since the only perspective from which we can look at the world is from inside our own brains. Given that an objective "view from nowhere" is not at our disposal, *perspectivalism* further extends this argument by emphasizing that even in the case of what, at first glance, appears to be an objective, context-independent assertion about the natural world—things like "there are nine planets in the solar system"—turns out to be, in Baghramian's view, "a statement made from a human perspective and . . . informed by human perceptions and conceptions." Accordingly, perspectivalism dictates that our judgments and decisions are constrained by "the position we occupy in time and space, as well as our interests and background knowledge." Since considerable experimental evidence now suggests that brain functions can be heavily influenced by context, relativism stands as a plausible theoretical framework from which to seek a better understanding of the vagaries of the human mind, and the complex brain from which it emerges.

Relativism does not come naturally or easily to science. The Cartesian way of examining the world could not accommodate *any* form of relativism in its credo. Solidly founded on the belief that the newly coined scientific method endowed men with the ability to uncover universal facts and laws about nature, it in fact professed, according to Baghramian, that "the mind, i.e the inner, has the function of representing the outer—the mind-independent world." As we have seen, this full-hearted embrace of objective scientific truth and dominance over the fallible and subjective human senses and mind shaped the experimental approach that dominated twentieth-century neurophysiology. So, in addition to carving up the brain into the smallest possible building blocks of sensory interpretation, neuroscientists went out of their way to remove the unwelcome "confounding variables" created by that brain's own point of view—context-dependence and mind-dependence. From the measurements of the receptive fields of individual neurons and sensory maps embedded in different brain structures, neuroscientists tried to infer the way the brain represents, precisely as the Cartesians predicted, the minimalist replica of the real world created inside the lab.

Given the multiple intellectual earthquakes that shook the scientific world from the mid-nineteenth century to the early decades of the century that followed, Baghramian and John Barrow properly assert that plenty of fertile ground existed for a radical questioning of this assumed certainty in philosophical and scientific thinking. After all, in 1859, Charles Darwin's theory of evolution ripped out of the ground of credibility the last roots of Bible-based cosmology, and just five years later James Clerk Maxwell unveiled the electromagnetic nature of light and predicted that its velocity in a vacuum is a universal constant. Soon after, faith in the existence of absolute truth in nature, as well as the belief in man's ability to prove what truth there is, were both shaken to their foundations. Almost as if they had accepted leading roles in a play written by Nietzsche, first, in 1925, Werner Heisenberg and his uncertainty principle of quantum mechanics, which postulated that the better one measures the position (or momentum) of a given particle, the worse one will estimate the momentum (or position) of the same particle, pushed physics further into the realm of the very small, well beyond our daily perceptional abilities. Then, just a couple of years later, the Austrian mathematician Kurt Gödel's incompleteness theorem unsettled the orderly world of mathematics and logic by revealing that there are statements within arithmetic that are true but cannot be proven to be true. Moreover, the discovery of non-Euclidean geometry in the first half of the nineteenth century dislodged one of the most massive bedrocks of Cartesian thinking: Newton's theory of gravity. About sixty years after the German mathematician Bernhard Riemann's doctoral work amazed even his adviser, the formidable Johann Carl Friedrich Gauss of Göttingen, a four-dimensional space-time continuum of non-Euclidean geometry built upon Riemann's work offered the fundamental framework needed for a former Bern patent office clerk to reinvent the laws of physics yet again.

Undoubtedly, Albert Einstein's special and general theories of relativity represent the most unabridged success of relativistic thinking ever engendered by a human mind. In its special version, the theory of relativity proposes that, given that the velocity of light is constant in a vacuum, both space and time must be perceived differently by observers who are moving at a constant speed in relation to each other. Essentially, neither time nor space is absolute. Instead, they must be relativized in relation to the state of motion of a pair of observers, moving at constant speed in relation to each other. Such relativization of time and space accounts for a series of counterintuitive effects, things like time dilation, wherein observed time—in the classic example, two clocks carried by the pair of observers—goes out of sync, and for Lorentz's length contraction, wherein observed objects contract due to relative velocity. However, as the American physicist Brian Greene explains in his book, *The Elegant Universe*, one would have to be moving at a significant fraction of light speed to be able to document that a watch is registering much less time elapsed (time dilation), since the beginning of the trip, than the one left with a friend who stayed on Earth. By the same token, only at similarly huge speeds would an observer on Earth be able to demonstrate that the length of a spaceship seems to have been reduced significantly (Lorentz length contraction). These experiences are certainly not part of the relatively slow velocity of daily life. "Special relativity is not in our bones—we do not feel it," Greene writes. "Its implications are not a central part of our intuition."

Despite the near-full embrace of the theory of relativity, the concept of relativism remained highly contentious and controversial. Relativistic thinking generated an intense debate, involving highly contradictory views of what scientific inquiry really means. In this never-ending contest, in one corner Baghramian pits the argument that scientific knowledge is universal, since it can be verified anywhere at any time. For instance, according to Nobel Prize winner Sheldon Glashow, scientists "affirm that there are eternal, objective, extra historical, socially neutral, external and universal truths, and that the assemblage of these truths is what we call physical science. Natural laws can be discovered that are universal, invariable, inviolable, genderless and verifiable." Curiously enough, Glashow finishes his manifesto with, "This statement I cannot prove.... This is my faith." In the other corner she places Heisenberg, who—naturally—maintained a pose of uncertainty. "The aim of this research is no longer an understanding of atoms and their movements 'in themselves,'" he wrote in *The Physicist's Conception of Nature*. "From the very start we are involved in the argument between nature and man in which science plays only a part, so that the common division of the world into subject and object, inner world and outer world, body and soul, is no longer adequate and leads us into difficulties." Here, man confronts himself alone.

In this debate, as in many other scientific issues, I defer remorselessly to the views of another unsuspicious believer in the scientific method, the late Stephen Jay Gould. Despite not subscribing to the philosophical school of relativism, Gould argued that "our ways of learning about the world are strongly influenced by the social preconceptions and biased modes of thinking that each scientist must apply to any problem. The stereotype of a fully rational and objective 'scientific method' with individual scientists as logical (and interchangeable) robots is self-serving mythology." Instead, in Gould's view, "impartiality (even if desirable) is unattainable by human beings. . . . It is dangerous for a scholar even to imagine that he might attain complete neutrality for then one stops being vigilant about personal preferences and their influences—and then one truly falls victim to the dictates of prejudice. Objectivity must be operationally defined as fair treatment of data, not absence of preference." Here, the heavy weight of Gould's argument sails without difficulty through the tricky hurricane winds unleashed by Gödel's incompleteness theorem:

> Since all discovery emerges from an interaction of mind and nature, thoughtful scientists must scrutinize the many biases that record our socialization, our moment in political and geographic history, even the limitations (if we can hope to comprehend them from within) imposed by a mental machinery jury-rigged in the immensity of evolution.

In my particular definition of the brain's own point of view, the set of physiological constraints that natural evolution imposed on our brains plays the equivalent role that light has on the theory of relativity: it defines the universal biological constant around which our daily, brain-created models have to be relativized. Animal evolution in general, and the evolutionary history of mammals and primates in particular, has to be considered as the source of the constraints around which thinking revolves because the anatomical and functional organization of our brains has been sculpted by the process of natural evolution. Indeed, thanks to an unpredictable series of environmental events that unfolded across hundreds of millions of years, this process has yielded an optimal blueprint for the emergence of the type of primate brain that each of us enjoys, from the compact, convoluted arrangement of the human cortex, dictated by the necessity of limiting the size of a newborn's head so that it could slide through a mother's birth canal, to the mesh of connectivity of its billions of individual neurons, which communicate electrically at the mercy of metabolism, biochemistry, and physiology.

For example, the cerebral vascular system must weave through this huge mesh of neurons, limiting the amount of oxygen transported to the brain by red blood cells, and thus the oxidation process through which neuronal mitochondria produce adenosine triphosphate, the main energy delivery molecule in cells. For this reason, primate brains operate within the boundaries of a tight energy budget. Since electrical signaling through action potentials is very costly in terms of energy, our brains can only produce, as we have seen, a finite number of action potentials at each given moment in time to represent a particular type of message. Let's call this primary constraint on how our brain operates the *fixed energy budget constraint.*

There are many other such biological constraint factors. These biological limits mean that, despite the wondrous feats that it can achieve, there are finite and specific boundaries for what a human brain can do and how it can do it, restrictions defining the type and amount of raw information it can process and handle and the varieties of thinking, logic, and behaviors it can produce. In this context, ruptures from the concept of an absolute truth, such as those provoked by Heisenberg's uncertainty principle and Gödel's incompleteness theorem, may primarily point to the existence of mental barriers that the human brain may not be able to cross, a territory that will forever remain incomprehensible to our primate minds. That is, unless they are someday aided by a fabulous new tool that, created by the brain, helps its creator to surpass its own biological prison.

Obviously, natural evolution also defines the biological limits of the human body that the brain inhabits. These include not only the physical limits of our biological actuators, muscles, tendons, and bones, but also the precise range and sensitivity of the arrays of peripheral body sensory organs—true biological transducers—that sample information from the outside world (and from inside our bodies) to keep the central nervous system well informed. Due to the functional limits of our eyes, ears, skin, tongues, and noses, we see, hear, feel, taste, and smell just a small fraction of the world that exists out there. That explains why, contrary to special relativity, natural evolution is literally in our bones, as it is in everything else that defines human nature. This I define as the *body constraint.*

Because our collection of body sensors informs the brain about the current conditions of the outside world, the brain is also able to map the environmental constraints that limit the kinds of behaviors from which it can select to reach a particular goal. Yet, evolution also granted to the human brain access to a precious glimpse of past experience. Buried deep within our brains, there are traces of the statistics of a planet Earth that no longer exists, but continues to influence how our brains operate, since it helped to shape the range of neurophysiological strategies and behaviors that we use to guarantee the fulfillment of our most fundamental goals, such as surviving and reproducing—and extracting the most pleasure from the, alas, few brief interludes separating those two demanding tasks. As John Barrow writes, "Whether or not living things are aware of it, they are embodiments of theories about the laws of Nature drawn from the part of Nature that they have encountered." In chapter 9, I discussed how our brains have, in practical terms, stolen a great chunk of the control of their own future evolution from our genes by acquiring the capability of creating powerful tools, which work at all spatial scales of nature, and expand the reach of the human body. By combining this toolmaking capability with a lifelong potential for learning and adapting, the human brain has mastered the unique art of incorporating the very artifacts it creates as a seamless extension of the mental model of the body it inhabits.

This plastic capacity also endows the brain with the privilege of storing in its vast heaps of distributed memories the unique series of events that marked the progression of an individual existence. Such a preciously exclusive personal biography, a unique time-varying random walk through the vagaries of life, includes each of our individual encounters with the outside world, all social relationships we establish with other members of our species and many others, and our immersion and absorption into the prevailing culture and philosophy of our time. As such, our individual history, from birth to death, sculpts and constrains the collection of internal brain models. I refer to this variable as the *individual history record.*

The question then becomes, what is relativized around these three constraints to understand the world from the relativistic brain's own point of view? I propose that, like a soccer team formed by billions of mildly related players, given a set of fixed constraints and the mandate to produce a particular goal-oriented behavior under a particular brain-state and environmental context, the relativistic brain selects, out of a huge number of possibilities, groupings of spatiotemporal patterns of neural ensemble electrical activity that can accomplish the job at hand. Here, by the "spatial domain" I mean the three-dimensional mass of neurons (the ball in the wire and ball model) that, at any given moment in time, is recruited to achieve a goal, and the time dimension refers to the temporal distribution of spiking activity within this neuronal population. By relativizing this neuronal space-time continuum, the primate brain has found a way to constantly and optimally select viable solutions for what is known to be a typical inverse problem: that is, given an observed behavioral outcome, what finite combination of brain activity should one choose out of a gargantuan set of options to produce the desired outcome. In this case, the biggest issues are which neurons to recruit, from what parts of the brain, and what spatiotemporal firing patterns these neurons will produce. From an outside observer's point of view, a sequence of arm movements (or whatever action or behavior is triggered by the neuronal firing) may look pretty much identical. Yet, from the brain's own point of view, the neural ensemble firing patterns that generate this movement will be similar but never the same. Rather than being a passive and faithful painter of what the outside world looks like, as the Cartesians believed, the human brain actively exerts its own probabilistic point of view on everything upon which it puts its eyes or hands.

Furthermore, consider the definition of a complete perceptual experience, resulting from observing a scene outside a moving train. According to the relativistic brain hypothesis, what emerges in our minds, as a kind of mental movie of that scene, results from the product of a relentless collision of incoming multidimensional information, containing limited data samples from the external world, and the brain's own point of view created a priori as a result of a long and random history of previous encounters with similar scenes. This fateful collision forges the real taste and touch, as well as the meaning and emotions, associated with each and every one of the exquisite spectrum of human sensations and feelings experienced during our succinct conscious existence. This is why I said, in the previous chapter, that a relativistic brain starts seeing before it actually watches. Taken to its limits, this shift in reference raises some intriguing consequences. For instance, it challenges two of the scientific obsessions of our times: the quest to reproduce human consciousness through artificial intelligence and the claim that a so-called Theory of Everything that could compress all that exists in the cosmos into some manner of universal mathematical formalism.

The arguments favoring the emergence of a relativistic brain strongly suggest that the primate central nervous system, and the human mind in particular, may not be compressed into any type of classical computational algorithm. In other words, the human brain as a whole is simply noncomputable. As pointed out by Barrow, there is no equation that could ever generate things like beauty, pleasure, and good poetry, just to mention three examples from what is likely an infinite list. But even though I contend that a relativistic brain is not computable as a whole, Gödel's famous incompleteness theorem may allow us to compute plenty of intelligence with the electrical storms that emerge from subsets of neuronal space-time—enough, perhaps, to let an artificial device ascend into the realm of humanity. Yet, for that to happen, such a machine would have to resign itself to becoming an assimilated part of a brain model that defines a unique human self.

Of course, if the human mind is noncomputable, there seems to be little hope that theoretical physicists will be able to untangle a radically reductionist Theory of Everything from the deepest ten-dimensional realm of 10^{-33}-centimeter-long vibrating strings. What are the chances of doing so, when computing the masterpiece play manufactured by the Brazilian soccer team seems impossible already? Not even the great Pelé could have imagined that, with a lovely magic flick of his right foot, he would have demonstrated, as John Barrow professed, that "prospective properties are beyond the reach of mere technique. They are outside the grasp of any mathematical Theory of Everything. That is why no non-poetic account of reality can be complete."

Still, we may be able to explore the deeper consequences of having our ambitious relativistic brains freely interacting with machines, and perhaps among themselves, through more profound media than traditional spoken language or virtual chat rooms. What that future of such interfaces may bring to each individual person and for our whole species is the topic of the next chapter. There, I will freely speculate about what may happen in the future when about fourteen hundred grams of relativistic gray matter acquires the full-fledged power to liberate itself from its body prison and, after mingling with other sources of gray matter, decides to embark on a reunion trip around the celestial confines from which it was born.

If such a voyage ever happens, it would crown the long epic of improbable odds the human brain has endured thus far to forge concrete reality for the entirety of human evolutionary and personal history, having as raw material only billions of noisy, biophysically challenged, probabilistic neurons. From its humble, stardust seeds, and after millions of years evolving quietly on a modest but cozy water-splashed rock enchanted by the gravitational pull of a slowly decaying third-rate star, who could possibly imagine that, through natural evolution, the human brain could acquire the privilege to capture and hold the very relativistic core of the cosmos?

When this extraordinary time comes, when human brain activity can be freely broadcast to the universe, a few may argue that we inadvertently risk giving away the intimate secrets of our humanity to whoever out there cares to listen. I do not fear that. For whatever galaxies those waves of thoughts sail through, our audiences will be far more intrigued by the fact that when the moment finally came to create the human brain, it seems that our gods had no choice but to become master dice throwers.

13

BACK TO THE STARS

The ritual, although well known by its participants, never seemed to grow old as the years went by. Late each afternoon, when the lazy tropical sunset began to inhibit the children's outside play, I could not wait for the moment when, always in silence, Lygia would stroll graciously toward her favorite spot in the living room to become my secret and willing accomplice. Inside her handsome white house with the genuine Tupi-Guarani hammock slung across the second-floor balcony, tucked into an amiable cul-de-sac, in Moema, a southern suburb of São Paulo, Lygia was certain I would never even be late for our daily musical rendezvous.

She was right. More than anything, every day, as I passed through the front door of her house without announcing myself, what I really wanted to witness were her imposing steps and the plume of rosewater she left behind as she moved elegantly to the box piano she had befriended through life's happy and not-so-happy moments.

Lygia Maria Rocha Leão Laporta had always been a beautiful, charming woman, and even if aging had brought a single shock of white to her otherwise immaculate black hair, it had not diminished the brightness that emanated from her light green eyes. Her hands, though delicate, carried the purpose and wisdom of someone who over many decades had explored the infinite possible combinations of exquisite movements, each and every one of them carefully designed and obsessively rehearsed, first in her head, and later by her fingers, in order to translate a long sequence of written notes, mixed with intimate emotions and memories, into a personal expression of music composed by someone else's brain hundreds of years earlier.

Listening and playing music was a large part of Lygia's life after her retirement. The rest of her time was devoted to learning anything and everything that she could in a single, fleeting life. Lygia knew very well the preciousness of time. At thirty-eight years of age, after losing the love of her life, her husband Vicente Laporta, to a brain tumor, she found herself in charge of her entire family. Alone, she raised two daughters and supported her mother, father, and brother, while working as a career civil servant. Unbeknownst to many of her friends and relatives, during those hard years Lygia managed to keep the most cherished of Vicente's dreams alive: the technical commercial school he had founded in 1943 in São Paulo. Vicente's ambition, the one he described to his bride on the day they first met, was to create similar schools around the country, so that students who could not afford or could not gain access to the few universities that then existed in Brazil would have an opportunity to claim better jobs and a better life through education. Although Vicente did not live long enough to pursue his full ambitions, Lygia continued to carry them for as long, and as far, as she possibly could.

When Lygia retired from her job in the mid-1960s, I became her sole student, in an informal school that matched Vicente's aspirations in its own special way. During those years, she was my best friend, my teacher, my true love, the person I could trust unconditionally to explain everything that I did not comprehend. So it was that the first museum I ever visited was the Museu do Ipiranga, on the day in 1972 when we celebrated the 150th anniversary of Brazil's independence from Portugal. The first opera I ever heard was *La Bohème*, at São Paulo's Teatro Municipal. The first time I saw the placid, lazy waves of the Atlantic Ocean was in the harbor city of Santos. In each of these unforgettable events, it was Lygia's hand that carried me into a completely new world, full of adventure, magic, and captivating people. But nothing was comparable to what I learned in Lygia's neat office, where I boldly traveled to places far beyond, out there, where only Captain Kirk and Dr. Spock, the Robinson family, and Dr. Zachary Smith and his robot dared to go. In Lygia's office I discovered how men learned to fly, and then, still not satisfied, how they decided, as Jules Verne imagined, to venture into the vast and empty cosmos. When Neil Armstrong landed on the moon in that magic summer of 1969, Lygia and I were sitting together in front of a black-and-white TV, both trying hard not to laugh as the people around us asserted that the broadcast was a Hollywood hoax.

By far, however, my favorite experiences came when Lygia, full of intense passion and tenderness, kindly bowed to her soccer-muddy but loyal follower and took her seat at the humble piano.

Knowing that I was observing every minuscule gesture of her performance, she seemed to take great care in elaborating every hand movement, as if attempting to carve a memory to last a lifetime. Clearly, she succeeded, for even now I can remember the way she rested her two delicate but determined hands on the keyboard before she started each piece. Indeed, I can still recall the sensation I felt when time seemed to cease ticking and the air in the room voluntarily stilled in anticipation of the moment when the first few notes would blast from the piano, like a tropical thunderstorm exploding in the atmosphere.

Although Lygia's choice of music changed from day to day, she more often than not devoted that opening salvo to express her devotion to the illustrious Polish composer Frédéric Chopin. Today, more than forty years after the afternoon when Lygia played her last concerto for me, I cannot listen to the first few notes of Chopin's *Polonaise héroïque* without revisiting those evenings in my grandmother's living room, where I discovered that, apart from memory, learning was the greatest of our brain's gifts. Unfortunately, Chopin's music will also be forever associated in my brain with the memories of my first unexpected and stunning acquaintance with the irremediably devastating effects that an insidious neurological disorder, progressing covertly, can have on someone's life. For it was in that same living room on a summer afternoon, that instead of playing, Lygia inexplicably stared at the piano for a few minutes in silence, until she turned to me and, without shifting her delicate hands from the keyboard, let two rivers of tears run from her puzzled eyes to tell me that she could no longer remember the sequence of movements she had repeated daily for most of her conscious adult life. The memories of how to produce the music she loved had abandoned her mind.

Unknown to her and to all of us at the time, Lygia had been suffering for several years from a continuous series of very small stroke episodes that gradually and mercilessly destroyed most of the upper layers of her frontal and parietal cortices. These cortical strokes resulted from thousands of minuscule blood clots—in medical terms, "emboli"—clogging the small blood vessels in her brain. These emboli did not produce any clear symptom until the destruction of brain tissue reached a critical level around the time Lygia's concerts came to an abrupt end. For years thereafter, she experienced a gradual and inexorable decay in her fine motor skills and memory, both of which contributed to occasional bouts of severe depression and sudden episodes of frightening self-awareness, in which she stated, to the despair of all of us and herself, that the person who once was Lygia was no more.

As Lygia's lifetime of memories, desires, loves, plans, and dreams slowly but emphatically evanesced forever, first from her brain, then from her mind, she started to lose grasp of the threads of conscious contact that she had shared with the people and world surrounding her. The last time we embraced, I felt, for a moment, that she did not recognize me.

My grandmother Lygia had a long and productive life. She did many things and left many happy memories in the minds of people who knew her. In our last long-distance conversation, after a few minutes of small talk, she became aware that it was her loyal student on the other end of the line. Without wasting a second, she straightened her voice and shouted:

"You know what time it is? You are late again, boy!"

"For what, Lygia, for what?" I simply could not understand what she meant.

"For Chopin, my son. For Chopin."

Over the past three decades, almost every time one of my scientific manuscripts returned from the mandatory peer-review process, I had to cope with the inevitable recommendation that all scraps of speculative thinking about our ability to interface brains and machines should be removed from the paper. During those painful reckonings, I would fantasize about the day when I could rescue those speculative ideas, and liberate them for others to consider and contemplate. That opportunity has finally arrived.

Such an exercise is not trivial given that, during the time I have spent confronting the ultraconservative culture of academia, a number of science fiction writers and movie directors have speculated unreservedly, and at times overindulged in the excesses of their fertile imaginations. During 2009 alone, two Hollywood mega-productions, *Surrogates* and James Cameron's *Avatar*, portrayed the stereotype of science being used to control, harm, kill, and conquer people with their technological wizardry. In these movies, BMIs allowed human beings to live, love, and fight by proxy. Their full-body avatars were left to do the hard work of roaming the universe and, in some cases, seeking to annihilate a whole alien race, on behalf of their human masters. Similarly violence-minded pop-culture renditions, from *Firefox* to the *Matrix* trilogy, help to reinforce the fear and anxiety spread by "futurologists," who warn us that humanity's doomsday waits just around the corner, since a revolutionary generation of extremely smart machines are about to take over our planet and make slaves of us all.

Here, I want to present an alternative view. After working and thinking long and hard about the impact of BMIs, I see a future filled with blunt optimism and eager anticipation, rather than with gloom and calamity. Perhaps because so little about the true dimensions of this future can be conceived with certainty, I feel an intense calling to embrace the alluring opportunities that freeing our brains from the limits of our terrestrial bodies will bring to our species. In fact, I wonder how anyone could think otherwise, given the tremendous humanistic prospects that BMI research promises to unleash.

But before I get to my vision of this future, I would like to allay some of the concerns voiced about the prospect of supremely intelligent machines to emulate, surpass, and dominate the many gifts of the human mind. Although I do not doubt for a minute that very sophisticated forms of machine intelligence may emerge someday, there is one virtually insurmountable impediment that any creator of such machines will face: it is extremely unlikely that any bootstrapped computing routine will ever be able to capture the precise temporal sequence of historical contingencies, at either the personal or evolutional time scale, that conspired to generate the human brain. In his breathtaking book *Wonderful Life*, Stephen Jay Gould masterfully lays down the basis for this argument by proposing the thought experiment he dubbed "replaying the life tape." In his view, no matter how many billions of microprocessors, teraflops, and terabytes or how many millions of artificial nucleotides are at one's disposal, the humongous efforts to create artificial intelligence will fail miserably, if the main and only goal is to build a mind comparable to our own. Here's how the experiment would unfold, according to Gould:

> You press the rewind button and, making sure you thoroughly erase everything that actually happened, go back to any time and place in the past—say, to the seas of the Burgess Shale. Then, let the tape run again and see if the repetition looks at all like the original. If each replay strongly resembles life's actual pathway, then we must conclude that what really happened pretty much had to occur. But suppose that the experimental versions all yield sensible results strikingly different from the actual history of life? What could we then say about the predictability of self-conscious intelligence?

Then, Gould offers his favored prediction for the most probable outcome of the experiment:

> Any replay of the tape [of life] would lead evolution down a pathway radically different from the road actually taken. But the consequent differences in outcome do not imply that evolution is senseless, and without meaningful pattern; the divergent route of the replay would be just as interpretable, just as explainable *after* the fact, as the actual road. But the diversity of possible itineraries does demonstrate that eventual results cannot be predicted at the outset. Each step proceeds for cause, but no finale can be specified at the start. . . . Alter any early event, ever so slightly and without apparent importance at the time, and evolution cascades into a radically different channel.

The particular array of contingencies that determined the evolution of the human brain may never be revisited, ever again, anywhere in the universe. Silicon-based consciousness, if it ever emerges, will almost certainly manifest itself in ways that are very distinct from our human version. As such, it is easy to see that our peculiar history also cannot be compressed into any computational algorithm, dashing any hope that machines, computer programs, or artificial forms of life could be subjected to an identical roll of evolutionary pressures generated by any computer code or other man-made gizmo. Effectively, one could say that, as a fair quid pro quo for carrying history's bequest within its circuits, the brain has been afforded the ultimate immunity against attempts to mimic or reproduce its most intimate secrets and skills.

The shielding provided by historical contingencies does not, however, guarantee that advanced machines may not one day come to dominate or even decimate the human race. Yet, I would rank the probability of such an event ever taking place at a level much lower than a multitude of other more palpable catastrophes that might account for the collapse of humanity. Environmental destruction, pandemics, famine, nuclear war, climate change, lack of freshwater, another meteor collision, the depletion of the ozone layer, even an alien invasion clearly have a higher probability of occurring and determining the downfall of our species than a potential coup d'état by the machines. In the very unlikely cataclysmic event that playing the "tape of life" could wreak such a destiny upon us, we can at least rest assured that our silicon conquerors will never, as John Barrow asserted, be able to comprehend the immortal meaning of such human verses:

> "Fear, O Achilles, the wrath of heaven;
> think on your own father and have compassion upon me, who am the more pitiable,
> for I have steeled myself as no man yet has ever steeled himself before me,
> and have raised to my lips the hand of him who slew my son."

Thus spoke Priam, and the heart of Achilles yearned as he bethought him of his father.

■ ■ ■

Personally, I would rather discuss how in the future humanity might take full advantage of the talents of the relativistic brain—its ability to simulate reality and its avid appetite to assimilate artificial tools, both to bypass neurological damage and to augment our reach and perceptions. As in my experiments recording neural ensembles, time will be my trusty guide. I will therefore begin by describing the biomedical applications of BMIs that will likely emerge in the next ten to twenty years. I will then move to a more distant future, perhaps several decades from now, when BMIs will be more commonplace, allowing us to merge with computational and virtual tools, devices, and environments. I will end my speculative voyage with what the remote future may bring to our species, when our mental allegiance to mortal bodies becomes less and less determinant of our way of living. Though I will not go into the details of the possible neuroengineering my vision entails as I look farther into the future, I am confident that the technological solutions will be found to make it a reality.

In the next two decades, brain-machine-brain interfaces, built by connecting large chunks of our brains through a bidirectional link, may be able to restore aspects of humanity to those who have succumbed, as in the case of my grandmother Lygia and my mentor Dr. César Timo-Iaria, to devastating neurological diseases. Possibly within a decade or two, BMIs will likely begin to restore neurological functions to the millions of people who can no longer hear, see, touch, grasp, walk, or talk by themselves.

An international research consortium, the Walk Again Project, which I cofounded, offers a first glimpse of this future. Conceived a few years after Belle and Aurora demonstrated the feasibility of linking living brain tissue to a variety of artificial tools, the project aims to develop and implement the first BMI capable of restoring full body mobility to patients suffering from severe body paralysis, whether it resulted from traumatic lesions of the spinal cord or from a neurodegenerative disorder (see Fig. 13.1). To accomplish this lofty goal, we are engineering a neuroprosthetic device that will allow paralyzed patients to use a BMI to control the

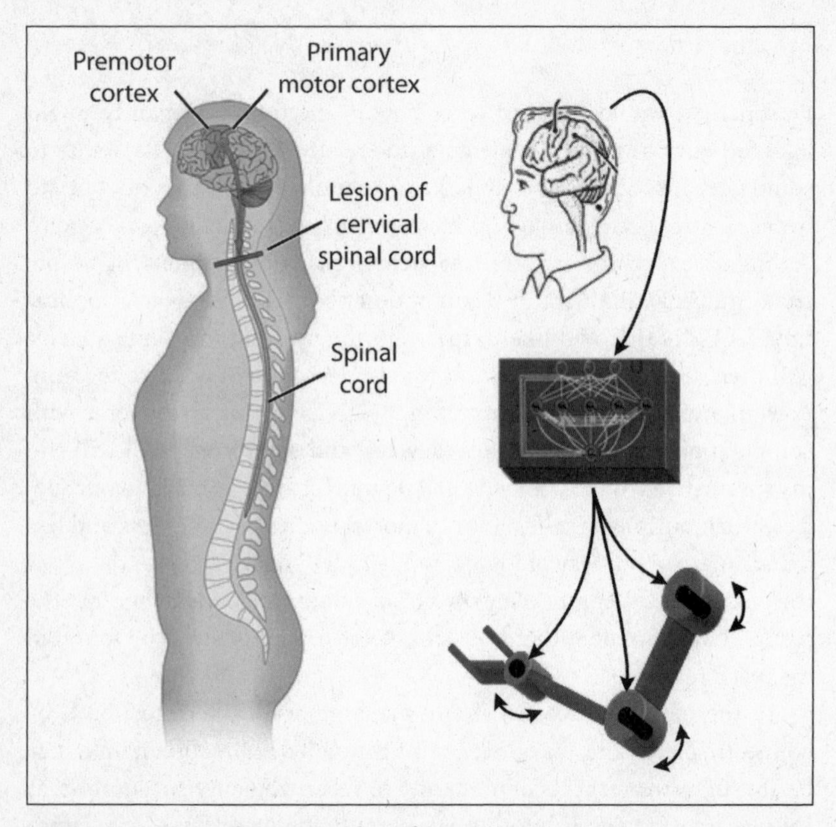

FIGURE 13.1 Cortical Neuroprosthesis for Restoring Motor Functions. A drawing illustrates how a cortical neuroprosthetic device may one day help patients paralyzed due to a spinal cord lesion (see text for details). *(Adapted from M.A.L. Nicolelis, "Brain-Machine Interfaces to Restore Motor Function and Probe Neural Circuits."* Nature Reviews Neuroscience *4 [2003]: 417–22.)*

movements of a full-body exoskeleton (see Fig. 13.2). This "wearable robot"—designed by none other than Gordon Cheng, now at the Technical University of Munich, the genial roboticist who made CB-1 learn how to walk under the control of Idoya's motor thoughts—will grant the patient control over his or her upper and lower limbs, and sustain and carry his or her body, as dictated by the person's voluntary will.

We are basing this feat of neuroengineering on the ten neurophysiological principles, derived empirically from our BMI experiments with Eshe, Aurora, Idoya, and many other animals, that have been described throughout this book. The BMI-controlled exoskeleton will require a new generation of high-density microelectrode cubes that can be safely implanted in the human brain and provide reliable, long-term simultaneous recordings of the electrical activity of tens of thousands of neurons, distributed across multiple brain locations. Indeed, to make BMIs clinically relevant and affordable, such large-scale brain activity recordings will have to remain stable for at least a decade without any need for surgical repair. Custom-designed neurochips will be chronically implanted in the skull, which will allow us to condition and process the brain's electrical patterns into signals capable of driving the exoskeleton. To reduce the risk of infection and damage to the cortex, these neurochips will also have to incorporate low-power, multichannel wireless technology, capable of transmitting the collective information generated by thousands of individual brain cells to a wearable processing unit, about the size of a modern cell phone. This unit will be responsible for running multiple,

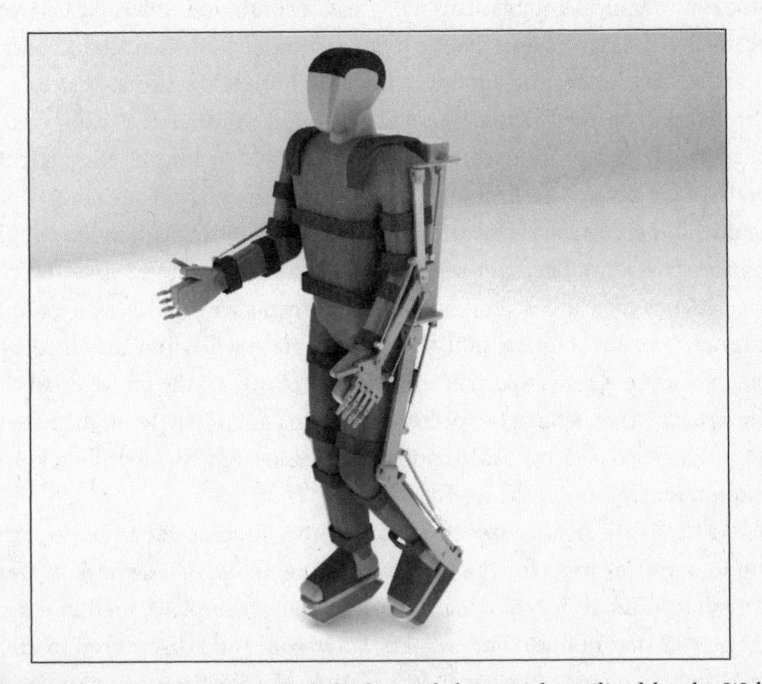

FIGURE 13.2 Design of the whole-body exoskeleton to be utilized by the Walk Again Project. *(Courtesy of Dr. Gordon Cheng, Technical University of Munich.)*

independent computational models designed to optimize the real-time extraction of motor parameters from the brain-derived electrical signals. It will also control all of the training programs employed as the patient learns how to operate the neuroprosthetic device.

The populations of neurons we sample to feed into this BMI will be distributed across multiple cortical and subcortical structures. Time-varying, kinematic, and dynamic digital motor signals will be extracted from this collective electrical activity of the brain and used to control the actuators distributed across the joints of the robotic exoskeleton. Following the current state-of-the-art techniques available for controlling such a device, high-order, brain-derived motor commands will interact with local electromechanical circuits, distributed across the exoskeleton, in order to mimic spinal-cord arc reflexes. This will permit the patient to initiate the walking step cycle, adjust gait speed, and trigger postural and gait adjustments in response to unexpected changes in the terrain. Meanwhile, low-level motor adjustments will be handled directly by the exoskeleton's electromechanical circuits. This will create a continuous interplay between brain-derived signals and robotic reflexes, a mode known as shared brain-machine control. I also envision force and stretch sensors, distributed throughout the exoskeleton, that will generate a continuous stream of artificial touch and proprioceptive feedback signals to inform the patient's brain of the device's performance. This array of signals will be delivered via multichannel cortical electrical microstimulation or through multiple light sources that stimulate light-sensitive ion channels deployed directly into the patient's cortex. Based on our lab experiments with BMIs, I expect that after a few weeks of interaction, the patient's brain will completely incorporate, via a process of experience-dependent plasticity, the entire exoskeleton as a true extension of the person's body image. At that point, the patient will be able to use the BMI-controlled exoskeleton to move freely and autonomously around the world.

As the Walk Again Project takes off, other lines of research are starting to show promise for the development of analogous devices to treat the symptoms of neurological disorders. For example, I mentioned in chapter 12 that Romulo Fuentes, Per Petersson, and I discovered in 2009 that high-frequency electrical stimulation of the surface of the spinal cord restored locomotion in both mice and rats that had been previously depleted of the neurotransmitter dopamine. Following this initial dopamine depletion, the rodents became severely rigid and immobile, exhibiting extreme difficulty in initiating any type of voluntary body movements—symptoms typical of Parkinson's disease. Because we could record the electrical activity of populations of neurons distributed across multiple cortical and subcortical structures in these animals, we observed that as the mice and rats became rigid, populations of neurons located in the motor cortex and the striatum started to fire in synchronous bursts of action potentials. All together, this synchronous activity produced an intense low-frequency neuronal oscillation that resembled an epileptic seizure (see Fig. 13.3). Interestingly, when we gave L-DOPA, the drug of choice to treat the early stages of Parkinson's disease, to the mice and rats, these oscillations were disrupted within a few minutes. Body rigidity gradually melted away, and the animals could move again. Once the effect of L-DOPA diminished, usually within a couple of hours, or once the animals developed a tolerance to the drug, usually within a few weeks, the rigidity returned with a vengeance.

Ten years earlier, Erika Fanselow and I had investigated methods for preventing epileptic seizures in rats. At that time, we had demonstrated that electrical stimulation of the trigeminal nerve disrupted the synchronous oscillations that marked the seizures we induced in our rats. This allowed the rats to escape from the behavioral arrest of an epileptic attack, and even precluded the onset of a new one. When I saw that the Parkinson's-like rigidity in the mice and rats was caused by synchronous oscillations that resembled those of an epileptic attack, I suggested to Fuentes and Petersson that we try the same approach with our dopamine-depleted rodents.

We first tried to electrically stimulate the trigeminal nerve. That produced some relief of facial rigidity, but no major effect on the rest of the body. What initially looked like a failure, in reality turned out to be the precious hint that we needed. Soon afterward, we switched the target of our electrical stimulation to the dorsal surface of the spinal cord. This new target had several advantages: it was much easier to reach, through a much less invasive surgical procedure, and it afforded us the opportunity to stimulate a huge number of large nerve fibers that run through that region on their way to the central nervous system, where they influence a

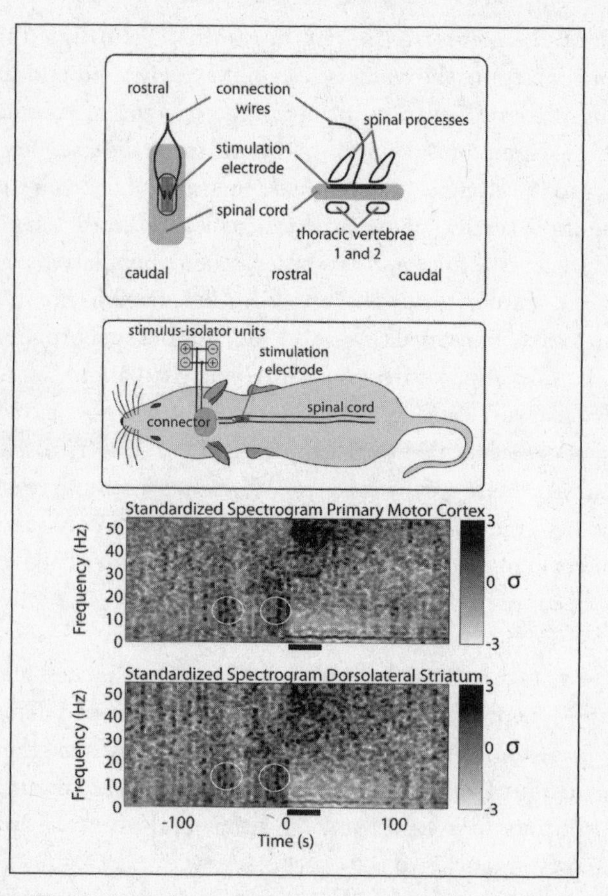

FIGURE 13.3 Treating a Parkinson's disease-like syndrome in rats using electrical stimulation of the spinal cord. On the top shelf, the stimulating electrodes and the implantation approach to place them on the dorsal surface of the spinal cord. Middle shelf shows an implanted rat that exhibits signs of Parkinson's disease. On the bottom shelf two circles are used to identify bursts of the epileptic activity observed in a spectrogram of the rat's brain activity, that are correlated with the akinesia produced by Parkinson's disease. At 0 time, the electrical stimulation of the spinal cord started, using the implanted electrodes. Notice that the epileptic activity disappears and, as a result, the rat was able to walk freely again (not shown). The spectrogram *X* axis represents time (0 start of stimulation) and the *Y* axis represents frequencies. The gray scale of the *Z* axis represents power or magnitude of brain activity at a given frequency (see scale on the right). *(Adapted from R. Fuentes, P. Petersson, W. B. Siesser, M. G. Caron, and M.A.L. Nicolelis, "Spinal Cord Stimulation Restores Locomotion in Animal Models of Parkinson's Disease."* Science *323 [2009]: 1578–82.)*

big chunk of the rat parietal and frontal cortices. It took only a brief trial to realize this was an optimal solution to the intense oscillations that were afflicting our mice and rats. As long as we kept the stimulation going, the rodents could roam around their cages, free from the rigidity produced by their Parkinson's-like disorder. Moreover, by continuously applying this stimulation, animals with a severe depletion of dopamine could be treated with a much smaller dose of L-DOPA. That lower dose reduced the side effects from the drug and the risk of developing a tolerance to it.

Unexpectedly, we had uncovered a very simple, minimally invasive, and inexpensive procedure—spinal cord stimulation—that may become the basis of a new therapeutic alternative for Parkinson's patients. Similar research is being feverishly conducted around the world in the hope of developing neuroprosthetic devices that interface with living brain tissue to treat other major neurological disorders, such as epilepsy and depression, and to restore vision, hearing, and speech.

In the near future, most BMI research will likely focus on the creation of novel therapeutic and rehabilitation tools. In all likelihood, however, the field will also contribute to a much deeper understanding of the neurophysiological principles that underlie the functioning of our relativistic brain in its efforts to compose and distort its model of reality.

I fully expect that BMI research will help to elucidate how the neuronal space-time continuum forms and operates, in a tight and cohesive way, throughout the course of our lives. To some degree, this issue pertains to the debate of the famous binding problem, described in chapter 4, a conundrum that has been haunting neuroscientists for quite some time now. By simply changing the frame of reference from incoming stimuli, generated by the outside world, to the vantage of the brain's own point of view, the binding problem might disappear altogether, since in the relativistic brain, there is no need to bind *anything*, because no incoming stimulus was broken into discrete sensory bits of information to begin with. In a relativistic brain there is simply a single dynamic model of the world that is continuously refreshed by the constant collisions between the brain's internal dynamics and the matching and nonmatching information sensed by the body's periphery.

Besides solving the binding problem, the relativistic brain theory may also offer a potentially viable truce for the intellectual war waged between the localizationist and distributionist camps of cortical physiology. I believe a compromise may be reached at long last if one accepts the notion that strict localization of functions in the cortex, as well as pure unimodal cortical representations, thrive only during the early development of the central nervous system or during reduced and artificial states of internal brain dynamics. We have seen how deep anesthesia, for instance, induces a collapse of internal brain dynamics, artificially limiting the complexity of individual neuronal sensory responses and whole cortical representations. Furthermore, it is only when animals actively engage in the exploration of their surrounding environment that the full dynamic splendor of the cortex is revealed. In the compromise I propose, discrete localization of functions and unimodal representation dominate the early postnatal developmental stages of the cortex, most likely because this is when the brain's connectivity is consolidated and the central nervous system gingerly crafts its internal models of reality. That gradual building up of the simulator and its models may account for the relatively long developmental periods of human childhood and adolescence. Indeed, this may explain why it takes several years for children to become capable of merging multimodal information describing the same object, such as the sound that is associated in their native language to the image of a corresponding letter or number.

At the end of this anatomofunctional maturation process, multiple multisynaptic cortical and subcortical loops connect populations of neurons across the cortex, giving rise to a single neuronal sea, ready to conduct the waves of neuronal time that make it tick. As the brain's own point of view evolves, the topographic maps, cortical structures, and strict cortical hierarchies become less dominant, until a point of no return is reached during early adulthood. By then, the brain tissue that still betrays a vestigial remnant of its anatomical modules—things like the cortical barrels of the rodent S1, ocular dominance columns, and cytoarchitectonic borders—speak primarily of times past, as indelible developmental scars of the organic scaffolding through which the brain had to crawl to assemble itself. Now these structures impose only minor constraints on how the neuronal space-time continuum functions.

Accordingly, to obtain a satisfactory comprehension of the mechanisms that generate a conscious and self-aware mind out of a blob of neural tissue, systems neuroscientists must shift their attention away from these mirages of the developmental past to follow more closely the waxing and waning waves and ripples of the brain's electrical ocean.

Such a newly acquired reliance on emergent internal brain dynamics, rather than strict allegiance to Brodmannian cortical geography, will, I believe, lead to a much more comprehensive understanding of neurological illness, since brain dynamics will become the medium through which we view all disturbances of the human mind. Neurological and psychiatric disorders will become linked to specific alterations in the timing of brain circuits and their interactions. If during its regular operation the brain experiences only delicate synchronous waves, during its altered states this neuronal sea may witness strange whirlpools and maelstroms of firing across the neuronal space-time continuum. Just as classical epilepsy is defined by a distinct pattern of synchronous brain activity, many other dysfunctions of the central nervous system may one day be ranked according to the level of pathological coherent firing in the brain. Considered from this perspective, the classical distinction between neurological and psychiatric disorders may simply vanish. As such, a better understanding of the principles of neural ensemble physiology will allow us to transcend both the ignominy of our collective ignorance about this group of particular mental states and the social stigma that surrounds those who live under their cloud. Ultimately, we may be able to recognize these disorders as what they truly are: mere disturbances of internal brain dynamics.

Preliminary support for this bold claim is already emerging in the work carried out by my former graduate student and postdoctoral fellow, Kafui Dzirasa, now an assistant professor of psychiatry at Duke's medical school. Dzirasa has systematically investigated variations in internal brain dynamics induced in a variety of transgenic mice, most of which have been created in Dr. Marc Caron's lab at Duke. In each one of these mice, a gene has been selectively removed. The mice have then been subjected to a sequence of pharmacological manipulations during adulthood. This allowed Dzirasa to engender in the mice a collection of stereotypical behaviors that resemble those observed in human patients suffering from a variety of cognitive dysfunctions and psychiatric disorders. By recording local field potentials and populations of single neurons in up to ten different brain structures in the mice, Dzirasa identified specific alterations in dynamic brain interactions that seem to be tightly correlated with the emergence of the abnormal phenotype expressed in each of the animals.

Although it is still difficult to establish a causal link between such neurophysiological alterations and expressed behaviors, Dzirasa has some striking evidence that may cross this pretty high threshold—including the identification of a potential neurophysiological basis for some of the stereotypical motor behaviors associated with obsessive-compulsive and bipolar syndromes. As in our experiments with dopamine-depleted rodents, the brains of these transgenic mice reveal differences in the level of coherent synchronous firing across the neuronal space-time continuum. Since we are now able to record large-scale brain activity for up to one year in these mice, we have documented, for the first time in the history of neuroscience, the progressive and inevitable neurophysiological transformation that an otherwise healthy brain undergoes when it takes a fateful turn toward one of the mental cul-de-sacs in which minds sometimes become stuck.

In the future, we expect to incorporate the information we are collecting in many strains of transgenic mice into a new framework, a more elaborate, multidimensional version of our Loch Ness state-space. This unabridged depiction of normal and altered dynamic brain states may allow us to classify most classical neurological and psychiatric disorders, similar to how we now associate different rat behaviors with distinct clusters on the three-dimensional state-space plot. In the long run, this research could allow neurologists to measure how a patient's brain dynamics behave and predict, well in advance of the appearance of any symptoms, whether there is a significant chance that a person may someday wake frozen by Parkinson's disease, depressed, or living in a completely new reality, dictated by full-blown mania, paranoia, or delirium. And by the same token, such a unified dynamic framework may allow physicians to evaluate quantitatively whether their treatments are working efficaciously and benefiting their patients.

■ ■ ■

The near-term prospects of basic and applied BMI and neuroscience research will entail an explosive acceleration in the convergence of disciplines ranging from computer science to biology, from engineering to medicine, and from mathematics to philosophy. As rising generations of neuroscientists employ a much broader bag of intellectual and technical tools, a multiplicity of transformative technologies will emerge. Traditional neuroscience departments and brain research institutes will have to adapt, in order to provide for the free interaction of experimental data, large-scale computer simulations, and theoretical work that will become the rule rather than the exception.

Already, many brain research collaborative initiatives have been established in an attempt to adapt to the neuroscience of the future. In fact, in March 2003, I launched a large-scale scientific endeavor of my own: the founding of a "Campus of the Brain," that today is known as the Edmond and Lily Safra International Institute of Neuroscience of Natal (ELS-IINN), a nonprofit academic effort located in the little town of Macaiba in the underdeveloped northeast coastal region of Brazil. The campus is dedicated to a threefold mission: to push the envelope of brain research to the limit; to celebrate the human brain's amazing achievements, manifested in terms of art, science, and culture; and to disseminate the resulting knowledge of the brain to the local society, through a series of social and economic projects—including a children's science education project, a women's health program, and an industrial research and technology park—aimed at uplifting the education, health, and living standards of the towns and communities close to the campus. Neuroscience as an agent of social transformation: I bet you never imagined that such a concept existed. One of the most ambitious projects will be the public school attached to the Campus of the Brain, in which children will be enrolled at the moment their pregnant mothers begin attending the prenatal care program offered by the campus's health center. By now, you may have guessed the name it will bear: the Lygia Maria Rocha Leao Laporta Public School.

In years to come, I sincerely hope the Brazilian Campus of the Brain becomes a model for establishing the types of multidisciplinary collaborations needed to actualize the future of brain-machine interfaces, since such scientific networks may greatly facilitate the receptivity to employing BMIs beyond the realm of rehabilitative medicine. Take, for example, what could happen within a few decades if we master technologies that allow humans to utilize the electrical activity of their brains to interact with all kinds of computational devices. From tiny personal computers that we carry with—or possibly within—us, to remote distributed networks aimed at mediating our digital social interactions, from the most mundane text processing to the most elaborate simulations of our secret dreams, our daily lives will look and feel significantly different from what we are accustomed to today.

For starters, interacting with the operating system and software of one's personal computer will likely become an embodied adventure, as our brain activity is used to grab virtual objects, trigger programs, write memos, and, above all, communicate freely with other members of our favorite brain-net, a considerably upgraded version of online social networking. The fact that Intel, Google, and Microsoft have already created their own brain-machine divisions shows that this idea is not far-fetched. The main obstacle is the need to develop a noninvasive method to sample the high-resolution brain activity needed to make such BMIs a reality. I feel confident that a solution will be found in the next couple of decades.

Then, what may sound unimaginable will become routine as augmented humans make their presence felt in a variety of remote environments, through avatars and artificial tools controlled by thought alone. From the depths of the oceans to the confines of supernovas, even to the tiny cracks of intracellular space inside our own bodies, human reach will finally catch up to our species' voracious ambitions to explore the unknown. It is in this context that I envision our brains will eventually complete their epic journey of emancipation from the obsolete terrestrial bodies they have inhabited for millions of years and utilize bidirectional, thought-driven interfaces to operate a myriad of nanotools that will serve as our new eyes, ears, and hands in the many tiny worlds crafted by nature. Worlds, made of clumps of atoms or balls of cells, into which our bodies could never have penetrated, but our thoughts certainly will, unopposed, unimpeded, and without any hesitation. Traveling in the opposite direction, we will likely be able to operate remotely controlled envoys and ambassadors, robots and airships of many shapes and sizes, sent on our behalf to explore other planets and stars in distant corners of the universe and capable of placing strange lands and scenery at the tip of our mental fingertips. With each step in our boundless explorations, the tools created by our descendants for these mind voyages will continue to be assimilated by their brains as further extensions of their selves, defining a brain's own point of view that is far, very far, beyond anything we can possibly imagine today. This thought, I must confess, brings me an enormous feeling of elation and awe, a profound emotion that, I presume, may resemble only something that a humble Portuguese sailor, five hundred years ago, may have experienced when, at the end of a long and life-threatening journey, he found himself staring at the bright sandy shores of a completely new world.

At this point in our speculative trip into the future, we might wonder how this tremendous expansion in human action and perception will affect the very rendering of reality that emanates from each of our progeny. Will they see and interpret the universe as we do, or will their daily experiences, ethics, culture, and science be so different from ours that a dialogue would be as impossible and meaningless as trying to debate the situation of the world's economy with a friendly group of Neanderthals today?

Ultimately, the most astounding consequence of liberating the human brain from the human body could be the unleashing of a powerful and unexpected set of contingencies capable of decisively influencing both the direction and the speed with which our species' "life tape" plays in the distant future. In other words, by the sheer power of what it can accomplish, emancipated from the constraints and vulnerabilities imposed by the human body, our relativistic brains may come to conquer the most coveted prize available to them in the universe: a hand in steering the evolution of our species.

Could such a complete liberation of the brain allow us to blur, or even eliminate, the once inexpugnable physical borders that define an individual human being? Could we one day, down the road of a remote future, experience what it is to be part of a conscious network of brains, a collectively thinking true brain-net? Assuming for a moment that somehow, through some amazing and harmless future technology, this brain-net became real, could the individuals participating in it not only communicate back and forth with one another just by thinking, but also vividly experience what their counterparts feel and perceive, as they seamlessly adhere to this true "mind meld"? Very few people today would likely choose to venture into these unknown waters, but it is impossible to know how future generations will react, given the opportunity to experience such a literally mind-boggling experience.

Accepting that all these stunning scenarios could actually take place, and taking for granted that such a collective mind meld became consensually accepted as an ethical way through which future generations interact and share their humanity, could these descendants of ours wake up one morning and simply realize that they had peacefully given birth to a different human species altogether? It is not inconceivable that our human progeny may indeed muster the skills, technology, and ethics needed to establish a functional brain-net, a medium through which billions of human beings consensually establish temporary direct contacts with fellow human beings through thought alone. What such a colossus of collective consciousness may look like, feel like, or do, neither I nor anyone in our present time can possibly conceive or utter. Like that goal of the Brazilian soccer team at the 1970 World Cup, this is the kind of greatness that one can only fully appreciate by experiencing its result as it unfolds, in all its complexity. It may, without our expecting it, proffer the ultimate human perceptual experience: to discover that each of us is not alone after all, that our most intimate thoughts, experiences, anguishes, passions, and desires, the very primordial stuff that defines us as humans, are shared by billions of our fellow brothers and sisters. The tremendous comfort that this might bring to so many, who feel like prisoners of their own haunting thoughts of isolation, inferiority, prejudice, misconception, and social inappropriateness, is difficult to imagine.

Although I am very aware that my particular optimistic viewpoint may not qualm all anxieties, I have no doubt that the rapacious voracity with which most of us share our lives on the Web today offers just a hint of the social hunger that resides deep in human nature. For this reason, if a brain-net ever becomes practicable, I suspect it will spread like a supernova explosion throughout the fabric of human societies. It's true that as individuals readily use their thoughts to control a huge range of artificial devices and communicate with one another, they may come to resemble something very unlike what we today call the human race. To this, I say: Given that our own species life's tape will continue to play its unpredictable tune, independent of what we think about the future, and that evolution will certainly not halt its course at some arbitrary stage down the road, why should we worry about who or what will succeed us thousands or even millions of years from now?

Having struggled with this questions for a long while, I believe the main reason we should worry about this future is neither grounded in any fear for our own particular destiny as a species, nor based on a reflexive repulsion to the idea that our species and way of life may, one day in the remote future, be replaced. Rather, I believe we owe to the safeguarding of the human heritage the same high standards of ethical and moral conduct we should devote, but unfortunately do not often do, to the preservation of every single form of life that inhabits our humble planet. From the stunning clouds of insects, communities of plants, squadrons of blue macaws, and packs of capybaras that roam through every cubic foot of the vast tropical rain forest, to the polar bears of the North Pole, to the spotted owls of western North America, and even to the last strains of the dreaded smallpox virus, preserving the diverse ways in which life manifests itself on our planet is our best way to pay tribute to the extraordinary circumstances that gave birth to the conscious mind. Preserving this biological patrimony is one of the first steps we can take to bestow a moral legacy to future generations, a legacy that by necessity must embrace not just every contextual trace, but every tiny bit of thinking, every imagined deed, good or bad, and each drop of the improbable neuronal elixir that bestows us with our sense of being.

How could we ever succeed in portraying the remarkable diversity of human experiences that compose the tale of our species' unique odyssey? Capriciously, the answer to this challenge may rest on the talents of our relativistic brain.

Back in 1945, the great Kurt Gödel stunned the scientific world, yet again, by proposing a new solution to Einstein's equations of general relativity. According to Gödel's solution, traveling back in time should be considered a distinct and real possibility in a relativistic universe governed by a space-time continuum and Riemannian geometry. Yet, despite being a mathematical possibility, traveling back in time would be far from trivial in practical terms and, as far as we can tell, is not the kind of experience that abounds in the universe—that is, unless you changed your frame of reference to another universe, another space-time continuum, the one inhabiting the space between your ears! There, within the confines of our inner neuronal universe, time travel becomes quite a trivial exercise; what any theoretical physicist would consider an astounding feat out in the space-time fabric of the stars, any of us can accomplish by merely swimming through the memories accumulated and carefully kept by the fabric of notes of our neuronal symphony, the waves that cross the space-time continuum of our minds.

If the future depicted in this chapter materializes, it takes just a minor leap of imagination to contrive that, in the midst of their newly acquired wisdom, our progeny may also decide to cross yet another Rubicon in our species' epic history and strive to document, for the benefit of future generations and the posterity of the cosmos, the richness and diversity of their human inheritance. Such an inestimable treasure could only be assembled, I suggest, by preserving the irreplaceable, first-person narrative of each and every single human lifetime story, the unique account of our mortal existence that, after a brief temporary stay in one's mind, is irremediably lost at the end of our lives, in a rare wasteful lapse of nature.

I can envision how a more thoughtful future society would concede to the download and storage of these lifetime chronicles, not only as a rite of passage at life's end, but also as a final tribute to yet another exceptional human life that inhabited this universe. Thereafter, each of these perennial records would be revered as a uniquely precious jewel, one among billions of equally exclusive minds that once lived, loved, suffered, and prospered, until they, too, became immortalized, not clad in cold and silent gravestones, but released through vivid thoughts, intensely lived loves, and mutually endured sorrows.

By then, the same wondrous technology and ethical covenant that will permit thoughts to be preserved forever will also allow them to be broadcast back to the edges of the universe, bringing, in the end, the ultimate sense of closure and solace to our kind that only a return to the maternal womb can afford. At this remote future juncture, I can still conceive one final eventful turn, as the fair and proper apotheosis for man's most improbable voyage through the contingencies of immemorial times: the crowning of our relativistic brain as the only meaningful trinity ever to bestow its blessings upon us. For, in addition to lodging the skillful sculptors of our sense of reality and self, and being the loyal guardians of our memories, it will be up to our brains, from there on, to effortlessly share, at light's speed, human symphonies, carefully composed over the course of a lifetime, with whoever and whatever are intrigued enough, anywhere in the vast cosmos, to just follow this music.

Sitting on a hill, at the construction site of the Campus of the Brain, as the ever bright equatorial sun prepares to take its daily well-earned rest, I can only wonder how someone, deep in our future, may one day react when he or she experiences, for the first time in the midst of his or her otherwise tranquil life as part of a collective consciousness, what it was to see, through mortal eyes made of flesh, a waving sea of palm trees, gently swaying back and forth, following the unrelenting brushing of the wind, as if they simply intended to blow a good-night kiss to a blossoming garden of cactus lying at their delicate, imperial feet. Perhaps, if listening with care, this distant cousin of ours would notice that the same wind, as it swirled through the concrete and steel foundations of the Lygia Maria Rocha Leao Laporta Public School, also seemed to whisper in my ears something I should already know: that I am late again. That it is time to stop playing in the muddy street and run, run as fast as I can, back to that house with the always-open door, and, once again, listen to Chopin.

SELECTED BIBLIOGRAPHY

1. WHAT IS THINKING?

Dawkins, Richard. *The Selfish Gene*. Oxford and New York: Oxford University Press, 1989.

Deutsch, David. *The Fabric of Reality: The Science of Parallel Universes—and Its Implications*. New York: Allen Lane, 1997, pp. 120–21.

Freeman, Walter J. *How Brains Make Up Their Minds*. New York: Columbia University Press, 2000.

———. *Mass Action in the Nervous System: Examination of the Neurophysiological Basis of Adaptive Behavior through the EEG*. New York: Academic Press, 1975.

Gaspari, Elio. *A Ditadura Envergonhada*, vol. 1, *Coleção As Ilusões Armadas*. São Paulo: Cia da Letras, 2002.

Hebb, Donald O. *The Organization of Behavior: A Neuropsychological Theory*. New York: Wiley, 1949.

Hubel, David H. *Eye, Brain, and Vision*. New York: Scientific American Library/W. H. Freeman, 1995.

Kauffman, Stuart A. *The Origins of Order: Self-Organization and Selection in Evolution*. New York: Oxford University Press, 1993.

Lashley, Karl. "In search of the engram." *Society of Experimental Biology Symposium* 4 (1950): 454–82.

———. *The Neuropsychology of Lashley: Selected Papers*, ed. by Frank A. Beach et al. New York: McGraw-Hill, 1960.

Mitchell, Melanie. *Complexity: A Guided Tour.* Oxford and New York: Oxford University Press, 2009.

Nicolelis, Miguel A. L., Gisela Tinone, Koichi Sameshima, et al. "Connection, a microcomputer program for storing and analyzing structural properties of neural circuits." *Computers and Biomedical Research* 23, no. 1 (1989): 64–81.

Nicolelis, Miguel A. L., Chia-Hong Yu, and Luiz Antonio Baccalá. "Structural characterization of the neural circuit responsible for the cardiovascular function control in high vertebrates." *Computers in Biology and Medicine* 20, no. 6 (1990): 379–400.

Sagan, Carl. *Cosmos.* New York: Random House, 1980, p. 4.

Shepherd, Gordon M. *Neurobiology*, 2nd ed. New York: Oxford University Press, 1988.

Weidman, Nadine M. *Constructing Scientific Psychology: Karl Lashley's Mind-Brain Debates.* New York: Cambridge University Press, 1999.

Zeki, Semir. *A Vision of the Brain.* Oxford and Boston: Blackwell Scientific Publications, 1993.

2. BRAINSTORM CHASERS

Adrian, Sir Edgar Douglas. *The Physical Background of Perception: The Waynflete Lectures Delivered in the College of St. Mary Magdalen, Oxford.* Oxford: Clarendon Press, 1947.

Broca, P. Paul. "Loss of speech, chronic softening and partial destruction of the anterior left lobe of the brain." First published in *Bulletin de la Société Anthropologique* 2 (1861): 235–38. Trans. by Christopher D. Green, York University, Toronto, Ontario, Canada, 2003, http://psychclassics.yorku.ca/Broca/perte-e.htm.

De Carlos, Juan A., and José Borrell. "A historical reflection of the contributions of Cajal and Golgi to the foundations of neuroscience." *Brain Research Reviews* 55 (2007): 8–16.

Erickson, Robert P. "The evolution and implications of population and modular neural coding ideas." *Progress in Brain Research* 130 (2001): 9–29.

———. "A study of the science of taste: On the origins and influence of core ideas." *Behavioral and Brain Studies* 31 (2008): 59–75.

Finger, Stanley. *Origins of Neuroscience: A History of Explorations into Brain Function.* New York: Oxford University Press, 1994.

Fritsch, Gustav, and Eduard Hitzig. "On the electrical excitability of the cerebrum" (1870). In *Some Papers on the Cerebral Cortex*, trans. by Gerhardt von Bonin (pp. 73–96). Springfield, Ill.: Thomas, 1960.

Gall, François [Franz] Joseph. *On the Functions of the Brain and of Each of Its Parts: With Observations on the Possibility of Determining the Instincts, Propensities, and Talents, or the Moral and Intellectual Dispositions of Men and Animals, by the Configuration of the Brain and Head*, 6 vols. Trans. by Winslow Lewis Jr. Boston: Marsh, Capen & Lyon, 1835.

Golgi, Camillo. "The neuron doctrine—theory and facts." Karolinska Institute, Stockholm, Sweden, December 11, 1906, http://nobelprize.org/nobel_prizes/medicine/laureates/1906/golgi-lecture.html.

Grant, Gunnar. "How the 1906 Nobel Prize in Physiology or Medicine was shared between Golgi and Cajal." *Brain Research Reviews* 55 (2007): 490–98.

Mörner, K. A. H. "Presentation speech." The Nobel Prize in Physiology or Medicine, Karolinska Institute, Stockholm, Sweden, December 10, 1906, http://nobelprize.org/nobel_prizes/medicine/laureates/1906/press.html.

Ramón y Cajal, Santiago. *Histology of the Nervous System of Man and Vertebrates*, vols. 1 and 2. Trans. by Neely Swanson and Larry W. Swanson. New York: Oxford University Press, 1995.

———. *Recollections of My Life*. Trans. by E. Horne Craigie and Juan Cano. Cambridge, Mass.: MIT Press, 1989.

———. "The structure and connexions of neurons." Karolinska Institute, Stockholm, Sweden, December 12, 1906, http://nobelprize.org/nobel_prizes/medicine/laureates/1906/cajal-lecture.html.

Robinson, Andrew. *The Last Man Who Knew Everything: Thomas Young, the Anonymous Polymath Who Proved Newton Wrong, Explained How We Can See, Cured the Sick, and Deciphered the Rosetta Stone, Among Other Feats of Genius*. New York: Pi Press, 2006.

Young, Thomas. *A Course of Lectures on Natural Philosophy and the Mechanical Arts*. London: Taylor and Walton, 1845.

———. The Bakerian Lecture: "On the theory of light and colours." *Philosophical Transactions of the Royal Society* 92 (1802): 12–48.

3. THE SIMULATED BODY

Blanke, Olaf, Christine Mohr, et al. "Linking out-of-body experience and self processing to mental own-body imagery at the temporoparietal junction." *Journal of Neuroscience* 25, no. 3 (2005): 550–57.

Blanke, Olaf, Stephanie Ortigue, et al. "Stimulating illusory own-body perceptions." *Nature* 419 (2002): 269.

Botvinick, Matthew, and Jonathan Cohen. "Rubber hands feel touch that eyes see." *Nature* 391 (1998): 756.

Brodie, Eric E., Anne Whyte, and Catherine A. Niven. "Analgesia through the looking-glass? A randomized controlled trial investigating the effect of viewing a 'virtual' limb upon phantom limb pain, sensation and movement." *European Journal of Pain* 11, no. 4 (2007): 428–36.

Brodmann, Korbinian. *Localisation in the Cerebral Cortex* (1909). Trans. by Laurence Garey. London: Smith-Gordon, 1994.

Ehrsson, Henrik, Birgitta Rosén, et al. "Upper limb amputees can be induced to experience a rubber hand as their own." *Brain* 131 (2008): 3443–52.

Herman, Joseph. "Phantom limb: From medical knowledge to folk wisdom and back." *Annals of Internal Medicine* 128, no. 1 (1998): 76–78.

Jasper, Herbert, and Wilder Penfield. *Epilepsy and the Functional Anatomy of the Human Brain*, 2nd ed. Boston: Little, Brown and Co., 1954.

Jeannerod, Marc. "The mechanism of self-recognition in humans." *Behavioural Brain Research* 142 (2003): 1–15.

Kemper, Thomas Le Brun, and Albert M. Galaburda. "Principles of cytoarchitectonics." In *Cerebral Cortex*, ed. Alan Peters and Edward Jones, vol. 1 (pp. 35–57). New York: Plenum Press, 1984.

Leyton, Albert S. F., and Charles Scott Sherrington. "Observations on the excitable cortex of the chimpanzee, orang-utan and gorilla." *Quarterly Journal of Experimental Psychology* 11 (1917): 135–222.

Makin, Tamar R., Nicholas P. Holmes, and H. Henrik Ehrsson. "On the other hand: Dummy hands and peripersonal space." *Behavioural Brain Research* 191 (2008): 1–10.

Melzack, Ronald. "From the gate to the neuromatrix." *Pain*, suppl. no. 6 (1999): S121–26.

———. "Phantom limbs." *Scientific American* 266, no. 4 (1992): 120–26.

———. *The Puzzle of Pain*. New York: Basic Books, 1973.

Melzack, Ronald, and Patrick D. Wall. "Pain mechanisms: A new theory." *Science* 150, no. 3699 (1965): 971–79.

Merzenich, Michael, Jon Kaas, et al. "Progression of change following median nerve section in the cortical representation of the hand in areas 3b and 1 in adult owl and squirrel monkeys." *Neuroscience* 10, no. 3 (1983): 639–65.

———. "Topographic reorganization of somatosensory cortical areas 3b and 1 in adult monkeys following restricted deafferentation." *Neuroscience* 8, no. 1 (1983): 33–55.

Murray, Craig, Stephen Pettifer, et al. "The treatment of phantom limb pain using immersive virtual reality: Three case studies." *Disability & Rehabilitation* 29, no. 18 (2007): 1465–69.

Nicolelis, Miguel A. L. "Living with ghostly limbs." *Scientific American Mind* 18 (2007): 53–59.

Nicolelis, Miguel A. L., Rick C. S. Lin, et al. "Peripheral block of ascending cutaneous information induces immediate spatiotemporal changes in thalamic networks." *Nature* 361 (1993): 533–36.

Penfield, Wilder, and Edwin Boldrey. "Somatic motor and sensory representation in the cerebral cortex of man as studied by electrical stimulation." *Brain* 60 (1937): 389–443.

Penfield, Wilder, and Theodore Rasmussen. *The Cerebral Cortex of Man: A Clinical Study of Localization of Function.* New York: Hafner Publishing Company, 1950.

Petkova, Valeria I., and H. Henrik Ehrsson. "If I were you: Perceptual illusion of body swapping." *PLoS ONE* 3, no. 12 (2008): e3832.

Pons, Tim, Preston E. Garraghty, et al. "Massive cortical reorganization after sensory deafferentation in adult macaques." *Science* 252, no. 5014 (1991): 1857–60.

Ramachandran, V. S., and Sandra Blakeslee. *Phantoms in the Brain: Proving the Mysteries of the Human Mind.* New York: William Morrow, 1998.

Wall, Patrick D. *Pain: The Science of Suffering.* New York: Columbia University Press, 2000.

Wall, Patrick D., and Ronald Melzack, eds. *Textbook of Pain*, 4th ed. Edinburgh and New York: Churchill Livingstone, 1999.

4. LISTENING TO THE CEREBRAL SYMPHONY

Berger, Hans. "Über das Elektrenkephalogramm des Menschen." *Archiv für Psychiatrie und Nervenkrankheiten* 87 (1929): 527–70.

Churchland, Patricia Smith, and Terrence J. Sejnowski. *The Computational Brain.* Cambridge, Mass.: MIT Press, 1992.

Evarts, Edward V. "Effects of sleep and waking on spontaneous and evoked discharge of single units in visual cortex." *Federation Proceedings* 19, Suppl. no. 4 (1960): 828–37.

———. "A review of the neurophysiological effects of lysergic acid diethylamide (LSD) and other psychotomimetic agents." *Annals of the New York Academy of Sciences* 66 (1957): 479–95.

———. "Temporal patterns of discharge of pyramidal tract neurons during sleep and waking in the monkey." *Journal of Neurophysiology* 27, no. 2 (1964): 152–71.

Hubel, David, and Torsten Wiesel. "Receptive fields, binocular interaction and functional architecture in the cat's visual cortex." *Journal of Physiology* 160 (1962): 106–54.

Lilly, John C. "Correlations between neurophysiological activity in the cortex and short-term behavior in the monkey." In *Biological and Biochemical Bases of Behavior*, ed. Harry F. Harlow and Clinton N. Woolsey (pp. 83–100). Madison: University of Wisconsin Press, 1958.

———. "Instantaneous relations between the activities of closely spaced zones on the cerebral cortex: Electrical figures during responses and spontaneous activity." *American Journal of Physiology* 176 (1954): 493–504.

Lilly, John C., George M. Austin, and William W. Chambers. "Threshold movements produced by excitation of cerebral cortex and efferent fibers with some parametric regions of rectangular current pulses (cats and monkeys)." *Journal of Neurophysiology* 15, no. 4 (1952): 319–41.

Lilly, John C., and Ruth B. Cherry. "Surface movements of the click responses from acoustic cerebral cortex of cat: Leading and trailing edges of a response figure." *Journal of Neurophysiology* 17, no. 6 (1954): 521–32.

McIlwain, James T. "Population coding: A historical sketch." *Progress in Brain Research* 120 (2001): 3–7.

Mountcastle, Vernon B. "Modality and topographic properties of single neurons of cat's somatic sensory cortex." *Journal of Neurophysiology* 20, no. 4 (1957): 408–34.

Niedermeyer, Ernst, and Fernando Lopes da Silva. *Electroencephalography: Basic Principles, Clinical Applications, and Related Fields*, 3rd ed. Baltimore: Williams & Williams, 1993.

Pauly, Philip J. "The political structure of the brain: Cerebral localization in Bismarckian Germany." *Electroneurobiología* 14, no. 1 (2005): 25–32.

Sherrington, Charles Scott. *Man on His Nature*, 2nd ed. Cambridge: Cambridge University Press, 1951.

Silk, Joseph. *The Big Bang: The Creation and Evolution of the Universe.* San Francisco: W. H. Freeman, 1980.

5. HOW RATS ESCAPE FROM CATS

Fox, Kevin. *Barrel Cortex.* Cambridge and New York: Cambridge University Press, 2008.

Georgopoulos, Apostolos P., Andrew B. Schwartz, et al. "Neuronal population coding of movement direction." *Science* 233, no. 4771 (1986): 1416–19.

Ghazanfar, Asif A., Christopher R. Stambaugh, et al. "Encoding of tactile stimulus location by somatosensory thalamocortical ensembles." *Journal of Neuroscience* 20, no. 10 (2000): 3761–75.

Nicolelis, Miguel A. L., ed. *Methods for Neural Ensemble Recording*, 2nd ed. Boca Raton, Fla.: CRC Press/Taylor & Francis, 2007.

Nicolelis, Miguel A. L., Luiz Antonio Baccalá, et al. "Sensorimotor encoding by synchronous neural ensemble activity at multiple levels of the somatosensory system." *Science* 268, no. 5215 (1995): 1353–58.

Nicolelis, Miguel A. L., Asif A. Ghazanfar, et al. "Reconstructing the engram: Simultaneous, multisite, many single neuron recordings." *Neuron* 18, no. 4 (1997): 529–37.

Nicolelis, Miguel A. L., and Sidarta Ribeiro. "Seeking the neural code." *Scientific American* 295, no. 6 (2006): 70–77.

Welker, C. "Microelectrode delineation of fine grain somatotopic organization of SmI cerebral neocortex in albino rat." *Brain Research* 26, no. 2 (1971): 259–75.

6. FREEING AURORA'S BRAIN

Carmena, Jose M., Mikhail A. Lebedev, Roy E. Crist, et al. "Learning to control a brain-machine interface for reaching and grasping by primates." *PLoS Biology* 1, no. 2 (2003): 193–208.

Carmena, Jose M., Mikhail A. Lebedev, Craig S. Henriquez, et al. "Stable ensemble performance with single-neuron variability during reaching movements in primates." *Journal of Neuroscience* 25, no. 46 (2005): 10712–16.

Chapin, John K., Karen A. Moxon, et al. "Real-time control of a robot arm using simultaneously recorded neurons in the motor cortex." *Nature Neuroscience* 2 (1999): 664–70.

Nicolelis, Miguel A. L., and John K. Chapin. "Controlling robots with the mind." *Scientific American* 287, no. 4 (2002): 24–31.

Nicolelis, Miguel A. L., Dragan Dimitrov, et al. "Chronic, multisite, multielectrode recordings in macaque monkeys." *Proceedings of the National Academy of Sciences* 100, no. 19 (2003): 11041–46.

Wessberg, Johan, Christopher R. Stambaugh, et al. "Real-time prediction of hand trajectory by ensembles of cortical neurons in primates." *Nature* 408 (2000): 361–65.

7. SELF-CONTROL

Fetz, Eberhard E. "Operant conditioning of cortical unit activity." *Science* 163, no. 870 (1969): 955–58.

Fetz, Eberhard E., and Dom V. Finocchio. "Operant conditioning of specific patterns of neural and muscular activity." *Science* 174, no. 7 (1971): 431–35.

Nowlis, David P., and Joe Kamiya. "The control of electroencephalographic alpha rhythms through auditory feedback and the associated mental activity." *Psychophysiology* 6, no. 4 (1970): 476–84.

Olds, James, and Marianne E. Olds. "Positive reinforcement produced by stimulating hypothalamus with iproniazid and other compounds." *Science* 127, no. 3307 (1958): 1175–76.

Schmidt, Edward M. "Single neuron recording from motor cortex as a possible source of signals for control of external devices." *Annals of Biomedical Engineering* 8 (1980): 339–49.

Wyricka, W., and M. Barry Sterman. "Instrumental conditioning of sensorimotor cortex EEG spindles in the waking cat." *Physiology & Behavior* 3 (1968): 703–7.

8. A MIND'S VOYAGE AROUND THE REAL WORLD

Birbaumer, Niels, Nimr Ghanayim, et al. "A spelling device for the paralysed." *Nature* 398 (1999): 297–98.

Birbaumer, Niels, Andrea Kubler, et al. "The thought translation device (TTD) for completely paralyzed patients." *IEEE Transactions on Rehabilitation Engineering* 8, no. 2 (2000): 190–93.

Blakeslee, Sandra. "Monkey's thoughts propel robot, a step that may help humans." *New York Times*, January 15, 2008.

Fitzsimmons, Nathan, Mikhail A. Lebedev, et al. "Extracting kinematic parameters for monkey bipedal walking from cortical neuronal ensemble activity." *Frontiers in Integrative Neuroscience* 3 (2009): 1–19.

Nicolelis, Miguel A. L. "Actions from thoughts." *Nature* 409 (2001): 403–7.

Patil, Parag G., Jose M. Carmena, et al. "Ensemble recordings of human subcortical neurons as a source of motor control signals for a brain-machine interface." *Neurosurgery* 55, no. 1 (2004): 27–35.

Peckham, P. Hunter, and Jayme S. Knutson. "Functional electrical stimulation for neuromuscular applications." *Annual Review of Biomedical Engineering* 7 (2005): 327–60.

Peikon, Ian D., Nathan Fitzsimmons, et al. "Three-dimensional, automated, real-time video system for tracking limb motion in brain-machine interface studies." *Journal of Neuroscience Methods* 180 (2009): 224–33.

Serruya, Mijail D., Nicholas G. Hatsopoulos, et al. [including John P. Donoghue]. "Instant neural control of a movement signal." *Nature* 416 (2002): 141–42.

Taylor, Dawn M., Stephen I. Helms Tillery, and Andrew B. Schwartz. "Direct cortical control of 3D neuroprosthetic devices." *Science* 296, no. 5574 (2002): 1829–32.

9. THE MAN WHOSE BODY WAS A PLANE

Berti, Anna, and Francesca Frassinetti. "When far becomes near: Re-mapping of space by tool use." *Journal of Cognitive Neuroscience* 12 (2000): 415–20.

Cardinali, Lucilla, Francesca Frassinetti, et al. "Tool-use induces morphological updating of the body schema." *Current Biology* 19, no. 12 (2009): R478–79.

Fisher, Helen E. *Why We Love: The Nature and Chemistry of Romantic Love.* New York: Henry Holt and Company, 2004.

Head, Henry, and Gordon Holmes. "Sensory disturbances from cerebral lesion." *Brain* 34 (1911): 102–254.

Hickok, Gregory, and David Poeppel. "The cortical organization of speech processing." *Nature Reviews Neuroscience* 8 (2007): 393–402.

Hoffman, Paul. *Wings of Madness: Alberto Santos-Dumont and the Invention of Flight.* New York: Hyperion, 2003.

Iriki, Atsushi, Masaaki Tanaka, et al. "Coding of modified body schema during tool use by macaque postcentral neurones." *Neuroreport* 7, no. 14 (1996): 2325–30.

Lebedev, Mikhail A., Jose M. Carmena, et al. "Cortical ensemble adaptation to represent velocity of an artificial actuator controlled by a brain-machine interface." *Journal of Neuroscience* 25, no. 19 (2005): 4681–93.

Maravita, Angelo, Charles Spence, et al. "Multisensory integration and the body schema: Close to hand and within reach." *Current Biology* 13, no. 13 (2003): r531–39.

Young, Larry J. "Being human: Love: neuroscience reveals all." *Nature* 457 (2009): 148.

Young, Larry J., and Zuoxin Wang. "The neurobiology of pair bonding." *Nature Neuroscience* 7, no. 10 (2004): 1048–54.

10. SHAPING AND SHARING MINDS

Chapin, John K., Karen A. Moxon, et al. "Real-time control of a robot arm using simultaneously recorded neurons in the motor cortex." *Nature Neuroscience* 2, no. 7 (1999): 664–70.

Delgado, José M. R. *Physical Control of the Mind: Toward a Psychocivilized Society.* New York: Harper & Row, 1969.

Fitzsimmons, Nathan, Weying Drake, et al. "Primate reaching cued by multichannel spatiotemporal cortical microstimulation." *Journal of Neuroscience* 27, no. 21 (2007): 5593–602.

Gell-Mann, Murray. *The Quark and the Jaguar: Adventures in the Simple and the Complex.* New York: W. H. Freeman, 1994.

Horgan, John. "The forgotten era of brain chips." *Scientific American* 293, no. 4 (2005): 66–73.

Nicolelis, Miguel A. L., and John K. Chapin. "Controlling robots with the mind." *Scientific American* 287, no. 4 (2002): 46–53.

O'Doherty, Joseph E., Mikhail A. Lebedev, et al. "A brain-machine interface instructed by direct intracortical microstimulation." *Frontiers in Integrative Neuroscience* 3 (2009): 1–10.

Serruya, Mijail D., Nicholas G. Hatsopoulos, et al. "Instant neural control of a movement signal." *Nature* 416, no. 6877 (2002): 141–42.

11. THE MONSTER HIDDEN IN THE BRAIN

Buzsáki, György. *Rhythms of the Brain.* Oxford and New York: Oxford University Press, 2006.

Dzirasa, Kafui, Sidarta Ribeiro, et al. "Dopaminergic control of sleep-wake states." *Journal of Neuroscience* 26, no. 41 (2006): 10577–89.

Gervasoni, Damien, Shih-Chieh Lin, et al. "Global forebrain dynamics predict rat behavioral states and their transitions." *Journal of Neuroscience* 24, no. 49 (2004): 11137–47.

Llinas, Rodolfo R. *I of the Vortex: From Neurons to Self.* Cambridge, Mass.: MIT Press, 2001.

Stapleton, Jennifer R., Michael L. Lavine, et al. "Rapid taste responses in the gustatory cortex during licking." *Journal of Neuroscience* 26, no. 15 (2006): 4126–38.

12. COMPUTING WITH A RELATIVISTIC BRAIN

Anokhin, Peter K. *Biology and Neurophysiology of the Conditioned Reflex and Its Role in Adaptive Behavior.* Trans. by Samuel A. Corson. Oxford and New York: Pergamon Press, 1974.

Baghramian, Maria. *Relativism.* London and New York: Routledge, 2004.

Barrow, John D. *Impossibility: The Limits of Science and the Science of Limits*. Oxford and New York: Oxford University Press, 1998.

Casti, John L., and Werner DePauli. *Gödel: A Life of Logic*. Cambridge, Mass.: Perseus, 2000.

Cohen, Leonardo G., Pablo Celnik, et al. "Functional relevance of cross-modal plasticity in blind humans." *Nature* 389 (1997): 180–83.

Egiazaryan, Galina G., and Konstantin V. Sudakov. "Theory of functional systems in the scientific school of P. K. Anokhin." *Journal of the History of the Neurosciences* 16, no. 1 (2007): 194–205.

Einstein, Albert. *Relativity: The Special and the General Theory*. N.p.: Quality Classics, 2009.

Frostig, Ron D., Ying Xiong, et al. "Large-scale organization of rat sensorimotor cortex based on a motif of large activation spreads." *Journal of Neuroscience* 28, no. 49 (2008): 13274–84.

Fuentes, Romulo, Per Petersson, et al. "Spinal cord stimulation restores locomotion in animal models of Parkinson's disease." *Science* 323, no. 5921 (2009): 1578–82.

Galeano, Eduardo H. *Soccer in Sun and Shadow*. London and New York: Verso, 1998.

Ghazanfar, Asif A., Chandramouli Chandrasekaran, and Nikos K. Logothetis. "Interactions between the superior temporal sulcus and auditory cortex mediate dynamic face/voice integration in rhesus monkeys." *Journal of Neuroscience* 28, no. 17 (2008): 4457–69.

Ghazanfar, Asif A., and Charles E. Schroeder. "Is neocortex essentially multisensory?" *Trends in Cognitive Sciences* 10, no. 6 (2006): 278–85.

Glashow, Sheldon. "We believe that the world is knowable." Presentation at *The End of Science?*, 25th annual Nobel Conference at Gustavus Adolphus College, Saint Peter, Minnesota, October 3–4, 1989, as quoted in Baghramian, *Relativism*.

Gould, Stephen Jay. *Full House: The Spread of Excellence from Plato to Darwin*. New York: Three Rivers Press, 2007.

Greene, Brian. *The Elegant Universe: Superstrings, Hidden Dimensons, and the Quest for the Ultimate Theory*. New York: W. W. Norton, 1999, p. 25.

Heisenberg, Werner. *The Physicist's Conception of Nature*. Trans. by Arnold J. Pomerans. Westport, Conn.: Greenwood Press, 1970, as quoted in Baghramian, *Relativism*.

Isaacson, Walter. *Einstein: His Life and Universe*. New York: Simon & Schuster, 2007.

Kelley, Patricia H. "Stephen Jay Gould's winnowing fork: Science, religion, and creationism." In *Stephen Jay Gould: Reflections on His View of Life*, ed. Warren D. Allmon, Patricia H. Kelley, and Robert M. Ross (pp. 171–188). New York: Oxford University Press, 2009.

Merabet, Lofti B., Joseph F. Rizzo, David C. Somers, and Alvaro Pascual-Leone. "What blindness can tell us about seeing again." *Nature Neuroscience* 6 (2005): 71–77.

Merabet, Lofti B., Jascha D. Swisher, et al. [including David C. Somers]. "Combined activation and deactivation of visual cortex during tactile sensory processing." *Journal of Neurophysiology* 97 (2007): 1633–41.

Nicolelis, Miguel A. L., and Mikhail A. Lebedev. "Principles of neural ensemble physiology underlying the operation of brain-machine interfaces." *Nature Reviews Neuroscience* 10 (2009): 530–40.

Perlmutter, Steve I., Marc A. Maier, and Eberhard E. Fetz. "Activity of spinal interneurons and their effects on forearm muscles during voluntary wrist movements in the monkey." *Journal of Neurophysiology* 80, no. 5 (1998): 2475–94.

Ribeiro, Sidarta, et al. "Neurophysiological basis of metamodal processing in primary sensory cortices." In press, 2010.

Sadato, Norihiro, Alvaro Pascual-Leone, et al. "Activation of the primary visual cortex by Braille reading in blind subjects." *Nature* 380 (1996): 526–28.

Timo-Iaria, César, Nubio Negrào, et al. "Phases and states of sleep in the rat." *Physiology & Behavior* 5, no. 9 (1970): 1057–62.

Zhou, Yong-Di, and Joaquín M. Fuster. "Somatosensory cell response to an auditory cue in a haptic memory task." *Behavioral Brain Research* 153, no. 2 (2004): 573–78.

13. BACK TO THE STARS

Barrow, John. *The Constants of Nature: The Numbers That Encoded the Deepest Secrets of the Universe*. New York: Vintage, 2009.

Dzirasa, Kafui, H. Westley Phillips, et al. "Noradrenergic control of cortico-striato-thalamic and mesolimbic cross-structural synchrony." *Journal of Neuroscience* 30, no. 18 (2010): 6387–97.

Dzirasa, Kafui, Amy J. Ramsey, et al. "Hyperdopaminergia and NMDA receptor hypofunction disrupt neural phase signaling." *Journal of Neuroscience* 29, no. 25 (2009): 8215–24.

Gould, Stephen Jay. *Wonderful Life: The Burgess Shale and the Nature of History*. New York: W. W. Norton, 1989, pp. 48, 50, 51.

Homer. *The Iliad of Homer.* Trans. by Samuel Butler. New York: E. P. Dutton & Company, 1923, p. 413.

Kurzweil, Ray. *The Singularity Is Near: When Humans Transcend Biology.* New York: Penguin, 2007.

Lebedev, Mikhail A., and Miguel A. L. Nicolelis. "Brain machine interfaces: Past, present and future." *Trends in Neuroscience* 29 (2006): 536–46.

Nicolelis, Miguel A. L. "Building the knowledge archipelago." *Scientific American* online, January 17, 2008, http://www.scientificamerican.com/article.cfm?id=building-the-knowledge-archipelago.

Pribram, Karl H. *Brain and Perception: Holonomy and Structure in Figural Processing.* Hillsdale, N.J.: Lawrence Erlbaum Associates, 1991.

Soares, Christine. "Building a future on science." *Scientific American* 298, no. 2 (2008): 72–77.

ACKNOWLEDGMENTS

During the past twenty-seven years many people have participated in the events and experiments described in this book. A few of them have been briefly mentioned here, but a large number of professors, mentors, students, colleagues, collaborators, and friends remained anonymous despite their profound contributions to my work as a scientist. Thus, first and foremost, I would like to thank all of them for allowing me the privilege of working with them or simply being part of their personal and intellectual lives. In particular, I would like to thank one of my postdoctoral mentors, Dr. Rick Lin, for introducing me to many aspects of neuroscience, and one of my scientific heroes, Dr. Jon H. Kaas, who by his innumerous acts of kindness and amazing intellectual deeds has defined the model of scientist I wanted to emulate.

Some of my former students and close collaborators were kind enough to read the original manuscript and made invaluable suggestions, criticisms, and warnings. I am, therefore, very grateful to Drs. Asif Ghazanfar, Marshall Shuler, Sidarta Ribeiro, and Mikhail Lebedev for their rich and insightful contributions. It goes without saying, though, that any remaining outrageous and unorthodox thoughts, comments, beliefs, or metaphors left in the manuscript are not their fault but, instead, should be credited to the author's vast list of sins. Another former student, Dr. Nathan Fitzsimmons, gave me the great pleasure of being in charge of all the illustrations of the book. Nathan's superb job served as a wonderful closing of his academic career and the starting point of new adventures in which he will certainly be as successful as he was as a Ph.D. student.

For the past two decades I have discovered, to my surprise, that my parents had hidden from my sister and me, their only natural children as far as we knew, that I had many true "brothers" spread around the world. This postnatal brotherhood has played an essential role in my scientific and personal life and, as is often the case, allowed me to learn many things about friendship and science that no university or school can ever teach. In this context, my two American "brothers," Drs. Alan Rudolph and Sidney "Sampras" Simon have been an integral part of my scientific and personal adventures for the past seventeen years, involving three continents and many unforgettable stories. Without their friendship and continuous encouragement and support, this book would have never been finished, simply because there would be no story to tell. I am also grateful to my Israeli brother, Dr. Idan Segev, who since we first met in the middle of nowhere in the Negev Desert and, soon afterward, during a stunning Bastille Day on the streets of Paris, has become my true benchmark of what a humanist should be and do. In the same rank, I would like very much to thank my Egyptian-Swiss brother, the great mathematician and philosopher of science Dr. Ronald Cicurel, who in our uncountable lunches over pizza napoletana and Diet Coke (regular Coke for him) at Da Carlo's restaurant in Lausanne, taught me more mathematics, physics, and philosophy than I ever dreamed to learn. Without Ronald's eagerness to teach and sheer generosity to share his immense intellect, I am afraid crucial parts of this book would have never made the light of the day.

By the same token, I am profoundly indebted to a series of key individuals whose kindness and dedication made it possible for this project to become a reality. First, I would like to thank Dr. James Levine, my literary agent, and all his wonderful colleagues at Levine and Greenberg in New York City, for their enormous kindness and support for the past two years. Without Jim's calm steering and generous experienced guidance to his rookie author, it is difficult to believe how this project could have succeeded. Similarly, I would like to profusely thank my editor, Robin Dennis, who diligently interacted with me, day in and day out, for almost two years to make sure that I could learn to write "short," and not "long" like my Brazilian ancestors, and go directly to the point without too many adjectives and acrimony, like my Latin American genes would prefer. Working with Robin, a real pro, was not only a thrilling pleasure, but the most important writing experience of my life. Thus, I thank her for all the generosity, diligence, and understanding that were required to edit and sharpen the original manuscript into a final product. Having learned so much from her, I like to think that the unusual experience of working with a Brazilian writer may also have left something Robin can remember. After all, following our dealings she has somehow morphed into a truly fanatical soccer fan, incapable of missing a single game of the 2010 World Cup. I am also very grateful for the wonderful support and encouragement I received from everyone at Henry Holt and Times Books, particularly from Paul Golob, who oversaw this project since its first days, and Serena Jones, who took over from Robin in the final stages of production of the book. Many thanks to Christine Soares and the editors of *Scientific American* for helping me in my initial attempts to write to a lay audience and for allowing me to reproduce here some excerpts of those articles.

Throughout my career I have had the privilege to work or collaborate in some of the most distinguished academic institutions in the world. I would like to thank my colleagues at my laboratory at the Department of Neurobiology and the Center for Neuroengineering at Duke University, where I have spent the past seventeen years of my career, as well as my friends and close collaborators in other countries, like Dr. Patrick Aebischer, Dr. Hannes Bleuler, and Solaiman Shokur, Tamina Sisoko, Jaime Ruiz at the Ecole Polytechnique Fédérate de Lausanne, Dr. Jean Rossier at the École Supérieure de Physique et de Chimie Industrielles, and Professor Henri Korn at the Institut Pasteur in Paris, and the entire staff of the Edmond and Lily Safra International Institute of Neuroscience of Natal and the Alberto Santos Dumont Association for the Advancement of Science in Brazil (AASDAP). In particular, I would like to thank Dora Montenegro, her teachers, and the 1,400 students who attend AASDAP's three science schools in Brazil for the inspiration and joy they have so freely and kindly offered me in the past four years. I am also indebted to all the anonymous American taxpayers and donors who through the National Institutes of Health and other federal and private agencies have helped to fund my research over the past twenty-two years. Many thanks also to my good Brazilian-Swiss friend, Pierre Landolt, and the Sandoz Family Foundation for their support. I am also grateful for the friendship and support of Mrs. Lily Safra.

Only two people have read the original and revised manuscript more times than Robin and me. The first of these beloved heroines was Neiva Cristina Paraschiva. For her thoughtful comments and continuous support, I thank her from the bottom of my heart. The second heroine, Susan Halkiotis, my guardian angel at Duke University for the past decade, has read every line of this book many times and provided the best proofreading and witty advice any author this side of the Milky Way could hope for. Nothing I could write or say would do justice to the dedication, competence, and sheer passion with which "Professor Halkiotis," as I like to call her, dived into this project. There is no doubt in my mind that without Susan's help I would have never fulfilled my task. For this and for the precious mutual friendship our families share, I hereby recognize my infinite debt to her.

For the past three decades I have been lucky enough to count always on the friendship, love, and support of Laura Oliveira and our three beloved sons, Pedro, Rafael, and Daniel Nicolelis. Without their presence in my life, nothing would have mattered. I dedicate my entire career as small tokens of gratitude for their understanding and personal sacrifice, without which I would have never had the strength and freedom to pursue my scientific and humanistic dreams to the very limits of my Brazilian imagination.

Natal, August 2010

INDEX

Page numbers in *italics* refer to illustrations.

ABOUT THE AUTHOR

MIGUEL A. L. NICOLELIS, M.D., Ph.D., is the Anne W. Deane Professor of Neuroscience at Duke University and founder of Duke's Center for Neuroengineering. His award-winning research has been published in *Nature*, *Science*, and other leading scientific journals, as well as in *Scientific American*, which named him one of the twenty most influential scientists in the world. A member of the French, Brazilian, and Pontifical academies of sciences, he lives in North Carolina. For more information, go to www.beyondboundariesnicolelis.net.